MEDICAL GENETICS

PEARLS OF WISDOM

Warren G. Sanger, Ph.D.

NOTE

The intent of MEDICAL GENETICS Pearls of Wisdom is to serve as a study aid to improve performance on a standardized examination. It is not intended to be a source text of the knowledge base of medicine or to serve as a reference in any way for the practice of clinical medicine. Neither Boston Medical Publishing Corporation nor the editors warrant that the information in this text is complete or accurate. The reader is encouraged to verify each answer in several references. All drug use indications and dosages must be verified by the reader before administration of any compound.

The editors would like to extend thanks to Terri Lair for her excellent managing and editorial support.

Art Director: Maryse Charette

This book was produced using Times and Symbols fonts and computer based graphics with Macintosh® computers

ISBN: 1-58409-030-8

DEDICATION

I dedicate this Pearls review book to my wife and son, Drs. Dixie and Travis Sanger, who have been a motivation for my accomplishments and to my parents who taught me certain genetics principles on the farm. I am also indebted to my colleague friends with whom I interact on a daily basis. I thank Kathy Olin for her work in assimilating these questions. Last, but not lease, I would also like to dedicate this book to the thousands of medical students, graduate students, house officers and fellows for having allowed me the opportunity to participate in their education.

Warren

EDITOR

Warren G. Sanger, Ph.D., FFACMG
Director, Human Genetics Laboratory
Professor, Pediatrics & Pathology/Microbiology
Hattie B. Munroe Center for Human Genetics
Munroe-Meyer Institute
University of Nebraska Medical Center
Omaha, NE

WE APPRECIATE YOUR COMMENTS!

We appreciate your opinion and encourage you to send us any suggestions or recommendations. Please let us know if you discover any errors, or if there is any way we can make Pearls of Wisdom more helpful to you. We are also interested in recruiting new authors and editors. Please call, write, fax, or e-mail. We look forward to hearing from you.

Send comments to:

Boston Medical Publishing Corporation
4780 Linden Street, Lincoln, NE, 68516

888-MBOARDS (626-2737)
402-484-6118
Fax: 402-484-6552
E-mail: bmp@emedicine.com
www.bmppearls.com

INTRODUCTION

Congratulations!! *Medical Genetics Pearls of Wisdom* will help you improve your knowledge base in Medical Genetics and will assist you in passing Genetics examinations and improve your board scores. A few words are appropriate in discussing intent, format, and use.

The primary intent of *Pearls* is to service as a rapid review of Medical Genetics Principles and to serve as a study aid to improve performance on Medical Genetics and Medical board examinations. With this intent in mind, the text is written in a rapid fire question/answer format. The student receives immediate gratification with a single correct answer. An emphasis has been placed on distilling trivia and key facts that are easily overlooked, quickly forgotten, but somehow seem to be needed on MCAT, medical board examinations and on genetics board examinations.

Many questions have answers without explanations. This format is utilized to enhance the ease of reading and rate of learning. It may be that when you read an answer you may think "why is that?". If this happens to you, go check! It has been shown that information learned, as a response to seeking an answer to a particular question, is much better retained than when passively read. Use your *Pearls* with your Medical Genetics text handy and open, or if you are reviewing on an airplane or in your fishing boat, mark questions for further investigation.

Pearls risks accuracy be aggressively pruning complex concepts down to the simplest facts. The dynamic knowledge base and practice of Medical Genetics is not like that in that new research and practice occasionally deviate from that which appear to represent the correct answer for test purposes. This text is designed to maximize your score in a test and you should refer to your mentors for direction on current practice.

Pearls is designed to be used, not just read. It is an interactive text. Use a 3" x 5" card to cover the answers and attempt all questions. A study method which we strongly recommend is oral group study, preferably over an extended meal or pitchers. The mechanics of this method are simple and no one ever appears stupid. One person holds *Pearls* with answers covered, and reads the question. Each person, including the reader says "check" when he or she has an answer in mind. After everyone has checked in, someone states his/her answer. If this answer is correct, go on to the next one. If not, another person states his/her answer or the answer can be read. Usually, the person who checks in first gets the first shot at stating an answer. Try it – it's kind of fun!

Pearls is also designed to be reused several times to allow, dare we use the word, memorization. Two check boxes are provided for any scheme of keeping track of questions answered correctly or incorrectly.

We welcome your comments, suggestions and criticisms. Great efforts have been made to verify these questions and answers. There will be answers we have provided that are at variance with the answer you would prefer. This is most often attributed to the variance between the original source and the source which you have chosen to use. Please make us aware of any errors you find. We hope to make continuous improvements in a second edition would greatly appreciate any input with regard to format, organization, content, presentation, or about specific questions.

Please write to: Warren G. Sanger, Ph.D.
                   Center for Human Genetics
                   985440 Nebraska Medical Center
                   Omaha, NE 68198-5440

Or e-mail:

Study hard and good luck on the Boards!

W.S.

TABLE OF CONTENTS

CLINICAL GENETICS

☐☐ **What percent of pediatric hospital admissions are due to genetic disease?**

40%.

☐☐ **What percent of adult hospital admissions are related to genetic disease?**

10%.

☐☐ **Complications of chorionic villi sampling can include:**

Possible bleeding, miscarriage, maternal contamination of cells and confined chorionic mosaicism.

☐☐ **Chorionic villi sampling is routinely performed at what stage of pregnancy?**

10-12 weeks gestational age (past LMP).

☐☐ **An advantage of chorionic villi sampling over that of amniocentesis is:**

Earlier results.

☐☐ **A disadvantage of chorionic villi sampling, compared to amniocentesis for prenatal diagnosis is:**

Greater chance for error because of maternal cell contamination, placental mosaicism and greater risk of miscarriage.

☐☐ **Genetic amniocentesis is generally performed at what gestational age?**

16-18 weeks gestation (past LMP).

☐☐ **Early amniocentesis is considered to be performed at what stage of pregnancy?**

Before 15 weeks gestational age (past LMP).

☐☐ **Levels of alpha-fetoprotein can be measured in fetal blood, amniotic fluid and maternal serum. Elevations of alpha-fetoprotein in these tissues most likely signify:**

Neural tube defect, ventral wall defect, fetal distress or incorrect gestational age.

☐☐ **Appropriate screening for persons of Eastern European ancestry is that of:**

Testing for the carrier status of Tay Sachs disease, Canavan disease and Gaucher disease.

❑❑ **The Black population has an increased risk for:**

Sickle Cell disease and Sickle Cell carrier state.

❑❑ **Persons of Mediterranean ancestry have a higher risk of being a heterozygote for:**

Beta thalassemia.

❑❑ **The genetic counseling approach which includes support for decision making, full disclosure of options but not paternalistic counseling is best termed as:**

Nondirective genetic counseling.

❑❑ **The identification of individuals with a genetic predisposition to a disease or a genotype that puts them at an increased risk of having a child with a genetic disease is termed:**

Genetic screening.

❑❑ **A low maternal serum alpha-fetoprotein, a low estriol, and high human chorionic gonadotrophin levels are associated with an increased risk for:**

Down syndrome.

❑❑ **The recurrence risk for a couple who has a previous child with Tetrology of Fallot is:**

Approximately 2-3%.

❑❑ **During what period of intrauterine development is the embryo most susceptible to malformation because of teratogens?**

4-6 weeks post conception.

❑❑ **Microcephaly, hepatosplenomegaly, cataracts and low birth weight are most characteristic of:**

Prenatal intrauterine infection.

❑❑ **A single localized abnormal formation of tissue that initiates a chain of subsequent events resulting in altered development is most characteristic of:**

A malformation sequence.

❑❑ **What percent of all <u>recognized</u> pregnancies results in spontaneous pregnancy losses?**

15-20%.

❑❑ **What is the most common etiology of first trimester spontaneous abortions?**

Chromosome abnormalities.

❑❑ **What is the most likely reason for a high frequency of certain autosomal dominant diseases which have low fitness?**

New mutations.

❑❑ **A prenatal distortion of normally formed structures is considered to be a:**

Deformation.

❑❑ **The prenatal <u>diagnostic</u> procedure which has been determined to have the lowest risk for miscarriages or fetal complications is that of:**

Amniocentesis (maternal hormone screening is a <u>screen</u> and not a diagnostic test).

❑❑ **An environmental or external agent which induces the development of a congenital abnormality is called a:**

Teratogen.

❑❑ **A birth defect which results from an extrinsic event acting upon an originally normal developmental process is termed a:**

Disruption.

❑❑ **What process involves family history procurement, a discussion of indications for prenatal diagnosis procedures, explanation of test limitations, and discussion of possible risks for procedures?**

Prenatal genetic counseling.

❑❑ **A young woman who is affected with phenylketonuria (PKU) is contemplating pregnancy. You should advise her that:**

Strict dietary control is required prior to and after conception to prevent birth defects in her child.

❑❑ **A 9-year-old male shows developmental delay, mental retardation, elongated facies, large simplified ears, and hyperextensibility of the joints. There is also a family history of the mother having 2 brothers who have similar problems in addition to having macroorchidism. The most likely consideration in your differential diagnosis is that of:**

Fragile-X.

❑❑ **The primary risks associated with advanced paternal age are that of:**

Autosomal dominant mutations.

❑❑ **The most likely consideration for your differential diagnosis in an adolescent male with small testicles, gynecomastia, and delayed secondary sexual characteristics would be that of:**

Klinefelter syndrome (47,XXY) or variant.

❑❑ **A newborn with microphthalmia, polydactyly, and cleft lip/palate has a significant risk for having:**

Trisomy 13 (Patau syndrome).

❑❑ **An adolescent female with ovarian dysgenesis and short stature should have chromosome studies to rule out:**

Turner syndrome (or other sex chromosome abnormalities).

❑❑ **Two syndromes which are associated with very similar deletions involving chromosome 15 are that of:**

Prader-Willi and Angelman syndromes.

❑❑ **A male with an X-linked recessive trait marries a woman whose father also has the same condition. What is the chance a child of these parents will also have the same condition.**

50%.

❑❑ **A patient with two homologous chromosomes from the same parent and none from the other is called:**

Uniparental disomy.

❑❑ **Double fertilization or retention of a polar body frequently results in a conceptus with:**

Triploidy.

❑❑ **Tall stature, aortic aneurysm, scoliosis, subluxation of the lenses and joint hypermobility are classic features of:**

Marfan syndrome.

❑❑ **Geneticists who do not tell patients what actions to take perform a type of genetic counseling which is known as:**

Nondirective genetic counseling.

❑❑ **Different gene abnormalities which result in similar phenotypes is known as:**

Genetic heterogeneity.

❏❏ **Supplementation to the diet of women of reproductive age with folic acid has been shown to be effective in the reduction of what fetal abnormalities?**

Neural tube defects and, most likely, congenital heart disease and cleft lip/palate.

❏❏ **Phenylketonuria is generally treated by:**

Careful dietary control with restriction of phenylalanine.

❏❏ **Collagen deficiencies can be found in which genetic diseases:**

Elhers Danlos, Stickler syndrome, Osteogenesis Imperfecta, and Marfan syndrome.

❏❏ **The differential expression of a gene which is dependent upon the sex of the parent from which it is inherited is referred to as:**

Genetic imprinting.

❏❏ **Maternal serum alpha-fetoprotein (MSAFP) is most useful as a screening test for:**

Open neural tube defects.

❏❏ **The terminology for centric fusion involving two acrocentric chromosomes is:**

Robertsonian translocation.

❏❏ **Prader-Willi, Angelman, Williams, DiGeorge, Smith-Magenis and Miller-Dieker syndromes are considered to be examples of:**

Chromosome microdeletion syndromes which can be detected or confirmed by FISH.

❏❏ **The recurrence risk for a couple who has a previous child with the trisomy 21 form of Down syndrome is:**

1-2%.

❏❏ **The theoretical and realistic risks for a young couple in which the female has a balanced form of a Robertsonian 14;21 translocation of having a term child with Down syndrome is:**

Theoretically 33%, realistically 15%.

❏❏ **The recurrence risk for a young couple in which their first child has Down syndrome and the father has a balanced Robertsonian t(14;21) translocation is:**

Theoretically 33%, realistically 3-5%.

❑❑ **The prenatal diagnostic procedure which provides chromosome results earlier than amniocentesis is:**

Chorionic Villi Sampling (CVS).

❑❑ **A 46,X,r(X) nomenclature indicates a:**

Female with a ring X chromosome and some clinical features of Turner syndrome.

❑❑ **A condition which does not always pose clinical signs in persons carrying the gene is that of:**

Reduced penetrance.

❑❑ **The risk for a couple, in which the husband has hemophilia A, having a pregnancy affected with this condition would be:**

0%, since daughters would be carriers and sons would inherit the Y chromosome and be normal.

❑❑ **Trisomies are the result of:**

Meiotic or mitotic nondisjunction.

❑❑ **Chromosomal mosaicism is the result of:**

Meiotic nondisjunction, followed by a mitotic nondisjunction or by a mitotic nondisjunction after a diploid conception.

❑❑ **Of the prenatal diagnostic procedures including amniocentesis, chorionic villi sampling, and chordocentesis, which diagnostic test provides the lowest risk for fetal loss?**

Amniocentesis.

❑❑ **Of amniocentesis, chorionic villi sampling and chordocentesis prenatal diagnostic procedures, which procedure is associated with the highest risk for fetal loss?**

Chordocentesis.

❑❑ **A point mutation of the SRY gene on Yp might give rise to:**

An external female phenotype or male infertility.

❑❑ **An infant with microcephaly, lowset ears, micrognathia, and rocker bottom feet most likely has:**

Edward syndrome (trisomy 18).

❐❐ **In maternal serum prenatal triple marker screening, a woman has a low serum alpha-fetoprotein, q low screen estriol and an elevated human chorionic gonadotropin (hCG). She has an increased risk for:**

Down syndrome.

❐❐ **The rationale for recommending prenatal diagnosis for women 35 years of age or older is that:**

The risk of autosomal trisomies is greater than risks of complications posed by an amniocentesis procedure.

❐❐ **A young woman in your practice indicated that she had an alcoholic binge prior to conception. On questioning her, you found out that she has not consumed any alcohol since the third day past conception. You would counsel her that she will most likely have a child with:**

No adverse effects because of the alcohol use (providing the history is accurate).

❐❐ **A couple has one son with a unilateral cleft lip and palate with no other family history of similar problems. The recurrence risk for the next child being affected is approximately:**

3-5%.

❐❐ **A child's blood type is "O". What is the blood type which neither parent can have for this to happen.**

AB.

❐❐ **Severe oligohydramnios quite often results in:**

Potter sequence.

❐❐ **The prenatal diagnostic procedure which can be performed as early as 10 weeks gestational age is that of:**

Chorionic Villi Sampling (CVS).

❐❐ **The frequency that 45,X fetuses are lost during the first trimester is:**

90-95%.

❐❐ **The procedure-related fetal loss rate for amniocentesis is:**

1/200 - 1/500.

❑❑ **A man has Marfan syndrome and his wife has Stickler syndrome (both are autosomal dominant disorders). What is the probability the first child will be free of both of these autosomal dominant traits?**

1 in 4.

❑❑ **A 46,XY, del(15)(q11q13) is the karyotype of a patient with:**

Prader-Willi syndrome or Angelman syndrome.

❑❑ **A newborn infant with a cat-like cry and microcephaly should have chromosome studies to rule out:**

"Cri du Chat" (deletion 5p).

❑❑ **The mode of inheritance for colorblindness is most often:**

X-linked recessive.

❑❑ **Congenital abnormalities are defined as:**

Being apparent at birth (genetic, environmental or both).

❑❑ **The only mucopolysaccharidosis condition which is due to an X-linked recessive pattern is:**

Hunter syndrome (MPS-II).

❑❑ **The observable properties of a person, as a result of genotype and environment, is termed:**

Phenotype.

❑❑ **The phenomenon of a single gene being responsible for a number of distinct phenotypic effects is:**

Pleiotropy.

❑❑ **More than the usual number of fingers or toes is:**

Polydactyly.

❑❑ **You have a newborn in your practice who has microcephaly, cataracts, low birth weight, and polydactyly. Which of these features would not be consistent with and intrauterine infection?**

Polydactyly.

❑❑ **Meningomyelocele, anencephaly, spina bifida occulta are all examples of:**

Neural tube defects.

❏❏ **A 15-year-old is referred to you with multiple café au lait spots, axillary freckling, scoliosis, and neurofibromas. What is the most likely diagnosis?**

Neurofibromatosis type I.

❏❏ **Of all recognized pregnancies, what percentage miscarry?**

15%.

❏❏ **In myotonic dystrophy, the phenomenon which results in increasing severity in subsequent generation is:**

Trinucleotide repeats causing "anticipation".

❏❏ **A duplication of a segment of 17p results in what syndrome?**

Charcot- Marie-Tooth.

❏❏ **In genetic counseling, the provision of prognosis, and recurrence risks, are dependent upon:**

Precise syndrome identification.

❏❏ **A disruption in fetal development which results in limb amputations and facial clefts is most likely due to:**

Amniotic bands.

❏❏ **The provision of patients and their families with an accurate assessment of the conditions in their family and explaining their risks for recurrence for a genetic condition is considered to be:**

Genetic counseling.

❏❏ **Your patient has one normal 4-year-old son but has had three subsequent first trimester pregnancy losses. An appropriate recommendation would be:**

Chromosome studies on her and her husband to rule out a balanced chromosome state which might lead to an unbalanced state in any future conceptions.

❏❏ **Your patient consults you regarding her observations that all sons and daughters of her and other maternal relatives have a disorder which affects muscles but this disorder has never been transmitted by males in the family. This is most likely due to:**

Mitochondrial inheritance.

❑❑ **You have a 6-year-old male in your practice who displays mental retardation, elongated facies, long simplified ears and macroorchidism. In consulting with the mother, she also has a brother with the same types of clinical findings. What is your most likely recommendation for diagnostic studies?**

DNA fragile X analysis.

❑❑ **Expansion of a CAG trinucleotide repeat on chromosome 4p is characteristic of:**

Huntington's disease.

❑❑ **You have a 16-year-old male in your practice with small testicles and underdeveloped secondary sex characteristics. A primary diagnostic consideration would be:**

Klinefelter syndrome for which chromosome studies are necessary for diagnosis.

❑❑ **An infant with microphthalmia, polydactyly and cleft lip/cleft palate is most likely due to what chromosomal syndrome?**

Trisomy 13.

❑❑ **Cleft lip/palate, obesity, schizophrenia, arteriosclerosis and spina bifida are due to what type of inheritance pattern:**

Multifactorial.

❑❑ **The mode of inheritance for Huntington's disease, achondroplasia, Marfan syndrome, and neurofibromatosis are due to what type of inheritance pattern.**

Autosomal dominant.

❑❑ **You have a 15-year-old female in your practice with ovarian dysgenesis, short stature, and delayed secondary sexual characteristics. What is the most likely diagnosis and what diagnostic test is indicated?**

Probable Turner syndrome - chromosome studies.

❑❑ **What is the most important reason to screen newborns for Sickle Cell disease?**

To reduce the morbidity and mortality of affected infants.

❑❑ **A small for gestational age fetus with congenital heart defects, syndactyly, and an extremely large placenta is characteristic of:**

Triploidy.

❏❏ **The phenomenon which involves random X chromosome inactivation in females is:**

Lyonization (X chromosome inactivation during early embryo development).

❏❏ **Genetic heterogeneity is caused by:**

Different mutations or alleles causing similar phenotypes.

❏❏ **Of the following chromosomal syndromes, including Turner syndrome, Cri-du-chat syndrome, Down syndrome, trisomy 13, trisomy 18, and Klinefelter syndrome, which is the most likely syndrome to <u>not</u> be clinically identified at birth?**

Klinefelter syndrome.

❏❏ **Neural tube defects can be best prevented by:**

Supplementation of folic acid to all women of reproductive age.

❏❏ **You have a 3-year-old in your practice who has pancreatic insufficiency, recurrent pneumonia, poor weight gain, and had meconium illius as a newborn. What is the most likely diagnostic consideration?**

Cystic fibrosis.

❏❏ **Osteogenesis Imperfecta, Stickler syndrome, and Ehlers Danlos syndromes are the result of what type of defect?**

Collagen abnormalities.

❏❏ **You consult with a couple in which the husband had a brother who died of Duchenne muscular dystrophy. Your patient is currently pregnant. You inform her that risks for her child having Duchenne muscular dystrophy is:**

Not at an increased risk because this is an X linked condition which her husband apparently does not have and, therefore, cannot transmit the condition.

❏❏ **The most common type of genetic disease at <u>conception</u> is:**

Chromosomal (reported to be as high as 50%).

❏❏ **A translocation between two acrocentric chromosomes with breakpoints near the centromere is termed:**

Robertsonian (centric fusion).

❏❏ **You have a 5-year-old child in your practice with joint stiffness, corneal clouding, hepatomegaly, coarsen facial features, and is profoundly developmentally delayed. A primary consideration in your diagnosis is that of:**

Hurler disease (mucopolysaccharidosis type I).

❑❏ **You have a woman in your practice with no symptoms of a dominant condition but has an affected father, maternal grandmother, and two of her three children have the same condition. The most likely explanation for this is that of:**

Reduced penetrance.

❑❏ **A point mutation of the SRY gene on Yp might account for the following phenotype.**

Internal or external female phenotype in a person with an XY karyotype.

❑❏ **You have a 25-year-old patient in your practice who recently had a miscarriage which was diagnosed as 45,X. She asks what the recurrent risks for a similar problem in future pregnancies. You inform her that:**

Her risk is not significantly increased over that of the general population.

❑❏ **A chromosome deletion which is generally associated with Prader Willi and Angelman syndrome is that of:**

del(15)(q11q13).

❑❏ **You have a young patient with lysosomal storage of glycogen, lactic acidosis, cardiac hypertrophy, and the inability to produce free glucose. A primary consideration in your diagnosis is:**

Glycogen storage disorder.

❑❏ **You have a 16-year-old patient in your practice who has a recent diagnosis of schizophrenia and she and her parents are wondering about the risks for such a disorder being present in your patient's brothers and sisters. You inform them that the risk is approximately:**

5%.

❑❏ **You see a 9-year-old delayed patient in your practice who has adenoma sebaceum, hypopigmentation of her back, and no malformations. A probable diagnosis is:**

Tuberous sclerosis.

❑❏ **A couple in your practice has a son and a daughter with the same genetic condition, without any family history of the condition. They ask for your explanation of the cause and recurrence risk for this condition. You explain that:**

The etiology for their children's condition is most likely autosomal recessive and the recurrence risk is 25%. A referral for genetic counseling would be appropriate.

❑❑ **Your 35 year-old patient presents for her first prenatal visit for her current pregnancy. Her brother and her 10 year-old son are developmentally delayed and have macroorchidism. Prenatal diagnostic considerations for your patient's current pregnancy include:**

Possible fragile-X because of family history and trisomy because of maternal age.

❑❑ **Your newborn patient has classic clinical features of Down syndrome. Why is it necessary to perform chromosome studies?**

Confirm diagnosis and to determine if this is a trisomic or translocation state for recurrence risk reasons.

❑❑ **A 6'8" high school basketball player visits your office because of back problems. You note that he has mild scoliosis, unusually long fingers and probable aortic regurgitation. A consideration in your differential diagnosis should include:**

Marfan syndrome (an autosomal dominant condition - should also inquire about family history).

❑❑ **Consanguinity poses the greatest risk for what type of Mendelian trait?**

Autosomal recessive and multifactorial.

❑❑ **Your patient has an autosomal dominant condition with 60% penetrance. What is the risk for your patient's first child having clinical signs of the disease?**

30% (50 x 60% = 30%).

❑❑ **Your 28 year-old patient sees you at 17 weeks gestation (by LMP) because of a high human chorionic gonadotrophin (hCG) and a low maternal serum alpha-fetoprotein (MSAFP). An ultrasound confirms the gestational age at 17 weeks, but reveals significant nuchal folds. What action is indicated and for what reasons?**

Prenatal diagnosis (amniocentesis) for chromosome abnormalities.

❑❑ **A young couple consults you because of the inability to conceive. Upon physical exam, you note that the male has underdeveloped secondary sexual characteristics and small testicles. What would you first consider in your diagnosis?**

Possible Klinefelter syndrome (47,XXY) - chromosome studies.

❑❑ **You have a newborn female patient with congenital lymphoedema. What genetic condition should be in your differential diagnosis?**

Turner syndrome - chromosome studies.

❑❑ **What is the most important reason for diagnosing Turner syndrome in the neonatal period?**

Referral for possible life-threatening cardiac defects (i.e. coarctation of the aorta). Also, for later management with hormonal therapy for growth and development of secondary sexual features.

❑❑ **A couple has three children, all with cystic fibrosis and they ask what the risk is for their next child having this condition. This risk is:**

25% (autosomal recessive).

❑❑ **A clinically normal couple has two children with clinically confirmed achondroplasia (an autosomal dominant condition). What is the most likely explanation?**

Gonadal mosaicism in one parent.

❑❑ **Your patient has a previous child with spina bifida. Your prenatal advise to reduce this risk for a subsequent child with a neural tube defect is:**

Daily folic acid supplementation before and after conception.

❑❑ **Your 15-year-old patient presents with short stature and primary amenorrhea. Chromosome studies should be performed to confirm or rule out:**

Turner syndrome or Turner variant.

❑❑ **Chromosome studies are performed on your 15-year-old patient because of primary amenorrhea and the results reveal 46,XY. What are your recommendations?**

Genetic counseling and gonadectomy because of significant increased risk of gonadoblastoma.

❑❑ **Your 12-year-old patient presents with mental retardation, short stature, a large appetite and obesity. What is a primary diagnostic consideration?**

Prader-Willi syndrome - recommend chromosome studies and, if uninformative, FISH and DNA studies for possible maternal uniparental disomy.

❑❑ **An infant presents with multiple "cafe au lait spots" and seizures. A primary diagnostic consideration is:**

Neurofibromatosis.

❏❏ **A 24-year-old patient reports that his father and grandfather had Huntington's disease. He has questions about his risk of developing the disease and about risks to his offspring. You would advise him that:**

He should seek genetic counseling - testing is available. His theoretical risk of developing HD is 50% and offspring have a 25% risk without further testing.

❏❏ **What is the most common reason for a first trimester pregnancy loss?**

Chromosome abnormality.

❏❏ **A 17-year-old young man consults you because of significant acne problems. Discussions with his mother reveal that he also has a history of behavioral problems and he is significantly taller than his parents and other young men his age. Your differential should include:**

XYY - chromosome studies should be performed.

❏❏ **A couple in your practice has two normally pigmented sons and a newborn son with albinism. What is the most likely mode of inheritance?**

Autosomal recessive.

❏❏ **A young couple consults you because of a newly diagnosed son with confirmed Duchenne muscular dystrophy. They inform you that they have just had a positive pregnancy test and are concerned about risks for this pregnancy being affected. Without further testing, the risks for this fetus being affected with Duchenne muscular dystrophy is (Duchenne muscular dystrophy is an X-linked recessive condition)?**

50% if male; 0% if female (50% chance of being a carrier).

❏❏ **A mother of a 9-year-old son consults you because her son has had excessive appetites, marked weight gain, and a newborn history of hypotonia. Upon physical exam, he is overweight, has microorchidism and is developmentally delayed. A primary consideration is:**

Prader-Will syndrome - cytogenetics and molecular cytogenetics studies should be performed.

❏❏ **A couple consults you about their risks for having a child with achondroplasia (an autosomal dominant condition) and it is obvious that both have the condition. Their risks would be:**

75% achondroplasia (homozygous may be lethal); if so, 50% affected; 25% normal.

❏❏ **A husband and wife both have hemophilia A (X-linked recessive). Their risk of having a child with this condition is:**

100% for both daughters and sons (since all three X chromosomes of the parents have the mutation).

❐❐ **A husband and wife both are albino (albinism is autosomal recessive) and have three children with normal pigmentation. How can this be explained?**

Parents have different alleles for albinism.

❐❐ **A couple is referred to you because of three early unexplained miscarriages. Your recommendation should be:**

Chromosome analysis on both to rule out or confirm a balanced chromosome rearrangement in one of them which would, if present, predispose to chromosomally unbalanced conceptions, and possible live births with multiple congenital anomalies.

❐❐ **A baby is born with multiple congenital anomalies. What is your first urgent laboratory test?**

Chromosome studies.

❐❐ **Advanced maternal age (over 35) is mostly associated with what type of genetic risk?**

Aneuploidy.

❐❐ **A 30-year-old couple consults you because the husband has a family history of Huntington disease (HD) and they are in the mid-trimester of pregnancy. The husband has been diagnosed as having symptoms of HD. What are the risks of their child having HD?**

50%, testing is available with appropriate counseling and protocols.

❐❐ **Among the following trisomies at conception, which is the <u>least</u> likely to be associated with livebirth - +13, +18, +21, +22?**

+22.

❐❐ **Your 26-year-old female patient reports a history of schizophrenia, mental disturbances, infertility and has had some learning difficulties. A possible chromosome reason might be:**

47,XXX.

❐❐ **When a female has a balanced X chromosome translocation with an autosome, which of her X chromosomes would be inactivated?**

The normal X, since the autosomal portion of the translocation must be active for survival.

❐❐ **Cutis aplasia is most commonly seen in which chromosomal trisomy?**

Trisomy 13.

❏❏ **You have an infant in your practice with clinical findings of Down syndrome. What is necessary before you can counsel the parents about risks of another pregnancy having the same condition?**

The chromosome complement of the infant (i.e., trisomy, translocation, mosaicism) as well as mother's age and family history.

❏❏ **You have a young couple in your practice in which the female has a balanced reciprocal translocation. You advise them that:**

They can produce chromosomally normal, balanced or unbalanced offspring, the risks of which depend upon chromosomes and breakpoints involved.

❏❏ **A newborn in your practice has microcephaly, congenital heart disease, dysmorphic facial features and bilateral cleft lip. The most likely general etiology is:**

Chromosome abnormality.

❏❏ **The most common chromosome abnormality associated with first trimester miscarriages is:**

45,X.

❏❏ **The most common trisomy which is associated with miscarriage before the third trimester is:**

Trisomy 16.

❏❏ **You have an 8-year-old in your practice with a history of hypotonia, hyperphagia, moderate mental retardation, and cytogenetic documentation of a deletion of chromosome 15q12. The diagnosis is:**

Prader-Willi syndrome.

❏❏ **Normal meiosis followed by a nondisjunctional event in mitosis results in:**

Mosaicism.

❏❏ **Death because of chromosomal imbalance is most frequent during which stage of human development?**

First trimester of pregnancy.

❏❏ **Why is it important to perform chromosome analysis on a child who has a clinical diagnosis of Turner syndrome?**

To rule out a 45,X/46,XY mosaicism because of the risk of testicular tissue leading to gonadoblastoma (malignant tumor), to counsel the family, to treat the child with growth hormones, and to treat the child with hormones to counteract the effects of delayed puberty.

❏❏ **The mode of inheritance characterized by males and females being affected through multiple generations is that of:**

Autosomal dominant inheritance.

❏❏ **Parents have a son and daughter with the same condition, with no other family history. The most likely mode of inheritance is that of:**

Autosomal recessive inheritance.

❏❏ **Two parents with normal stature have a child with achondroplasia, a form of dwarfism which is considered to be autosomal dominant. What is the most likely reason for this?**

New mutation or gonadal mosaicism.

❏❏ **Two siblings have the same autosomal dominant disease; however, one is much more severely affected than the other. This is a reflection of:**

Variable expressivity.

❏❏ **Two unaffected parents have a child with a recessive disease. The probability for an unaffected sibling of this child being a carrier for the recessive gene is:**

2/3 (because we know that he/she is not affected).

❏❏ **When a genetic disease is not expressed by persons who have the abnormal gene associated with the disease, this is a phenomenon referred to as:**

Increased penetrance.

❏❏ **A child with achondroplasia is born to two parents of normal stature. The affected individual with achondroplasia has a risk for offspring being affected of:**

50%.

❏❏ **Your pregnant patient has a father who is affected with hemophilia B, an X-linked abnormality. What is the chance that her expectant son will be affected with this disorder?**

50% (patient is an obligate carrier and 50% chance of transmission).

❏❏ **A couple has 4 children with the same autosomal recessive disorder and they are expecting their fifth child. What is the chance that this fifth child will be affected?**

25%.

❑❑ **If two individuals have the same autosomal dominant disease marry, what proportion of the offspring would have the disease?**

75%, unless the homozygous abnormal is lethal and then the risk in offspring being affected would be 50%.

❑❑ **A parent has a 21;21 Robertsonian translocation, which is balanced. The risk for having a child with Down syndrome is:**

100%.

❑❑ **The most likely mode of inheritance for male to male transmission of a trait is that of:**

Autosomal dominant inheritance (less likely Y-linked).

❑❑ **Jackie's father has hemophilia A, an X-linked recessive condition. What is her chance of being a carrier?**

100% (Jackie inherited the abnormal X from her father, otherwise she would have inherited the Y and she would have been named Jack).

❑❑ **In the general population, what percent of newborns will be born with a genetic disease which is defined by structural malformation, a chromosome abnormality, or a gene abnormality?**

5%.

❑❑ **Advanced maternal age is most associated with what type of genetic abnormalities?**

Chromosomal trisomies.

❑❑ **Advanced paternal age is most associated with what type of genetic abnormalities?**

New gene mutations.

❑❑ **The most likely mode of inheritance in the following pedigree is:**

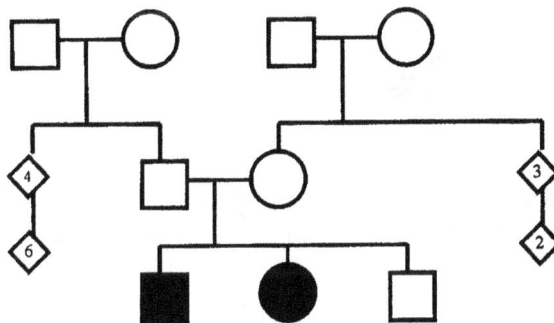

Autosomal recessive

❐❐ **The most likely mode of inheritance in the following pedigree is:**

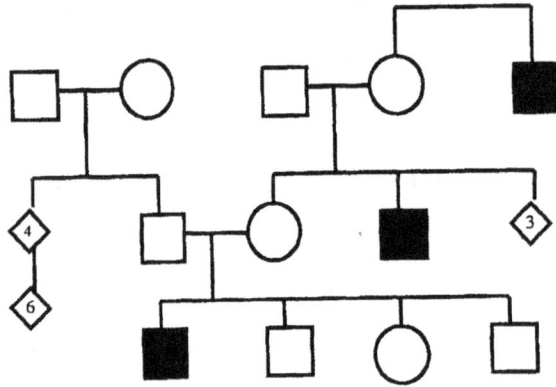

X-Linked recessive.

❐❐ **The most likely mode of inheritance in the following pedigree is:**

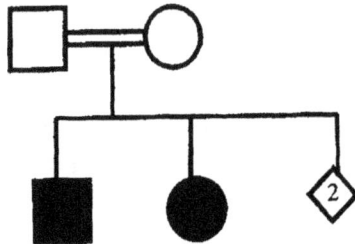

Autosomal recessive (consanguinity).

❐❐ **The most likely mode of inheritance in the following pedigree is:**

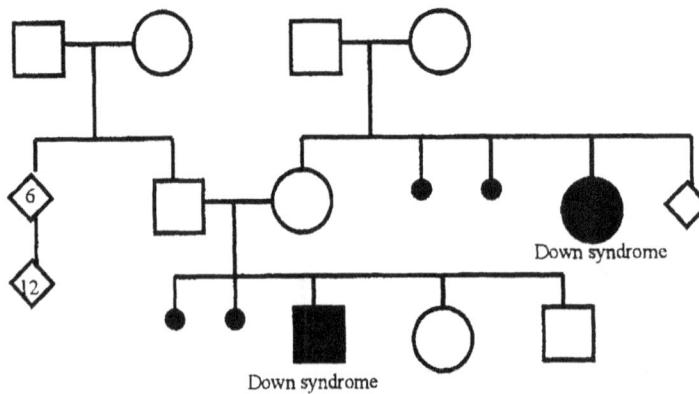

Maternal familial translocation involving chromosome 21 which is balanced in normal parents and unbalanced in Down syndrome offspring and miscarriages.

❏❏ **The most likely mode of inheritance in the following pedigree is:**

Y-linked or autosomal dominant with early female lethality.

❏❏ **The most likely mode of inheritance in the following pedigree is:**

Chromosome translocation, which is unbalanced in miscarriages and in infants with birth defects.

❏❏ **The ABO blood group is located on:**

Chromosome 9.

❏❏ **Achondroplasia, a form of dominant dwarfism, is located on chromosome:**

4.

❏❏ **You have an 8 year-old patient who has a cleft palate, tetralogy of Fallot, a height at the 4th percentile and mental retardation. What is the most likely diagnosis?**

Velocardiofacial syndrome.

❏❏ **You have a 5 year-old patient with cleft palate, a height at the 2nd percentile, myopia, a family history of retinal degeneration and detachment, and also has orthopedic problems. What is the most likely diagnosis?**

Stickler syndrome.

❏❏ **Huntington's disease generally has an onset in the mid-thirties. You have a 20-year-old patient who has presented with rigidity and a family history of Huntington's disease. What is the most likely reason for this?**

Paternal inheritance of the CAG-repeat expansion, also referred to anticipation.

❏❏ **A duplication of a gene on chromosome 17p is generally associated with:**

Charcot-Marie-Tooth syndrome.

❏❏ **Tay Sachs disease, Canavan disease, Bloom syndrome and Type I Gaucher disease is most frequently seen in what population?**

Ashkenazi Jewish population.

❏❏ **You have a couple who is being seen for genetic counseling because the husband's sister had a child with Tay Sachs disease. The husband has been tested and is found to be a carrier of Tay Sachs disease. There is no family history in the wife's family for Tay Sachs disease. The most appropriate next step would be:**

Perform carrier testing on the wife.

❏❏ **Velocardialfacial syndrome is most frequently associated with a deletion of chromosome:**

22q.

❏❏ **A chromosome deletion at chromosome 17p13 is most likely associated with a syndrome:**

Miller-Dieker syndrome.

❏❏ **You have a woman in your practice who asks what her risk is of having a subsequent child with a Rett syndrome, who has a previous daughter with Rett syndrome. You tell her the risk is:**

Below 5%, presuming the diagnosis is correct.

❏❏ **If genes on two homologs are exactly the same, this is termed:**

Homozygous.

❏❏ **If genes on the two homologs represent different alleles, this is termed:**

Heterozygous.

❏❏ **In approximately 5% of couples with 2 or more unexplained pregnancy losses, the etiology is:**

A balanced structural rearrangement or translocation in one of the parents.

❏❏ **A genetic counseling approach which involves providing information, support, but does not involve recommending a specific course of action is termed:**

Non-directive genetic counseling.

❏❏ **In a couple who has had two children with the same structural chromosome abnormality but both parents have normal peripheral blood chromosome studies, a likely etiology of the structural abnormality in their two children is:**

Gonadal mosaicism in one parent.

❏❏ **The most common syndrome known to be associated with maternal uniparental disomy is:**

Prader-Willi syndrome.

❏❏ **A risk term utilized for the intrinsic true probability or risk based upon previously collected data on similar family situations is termed:**

Empiric risk.

❏❏ **The two sex chromosome aneuploid situations which generally result in infertility are:**

47,XXY and 45,X.

❏❏ **Two common sex chromosome aneuploidy states which have little effect on fertility include:**

47,XXX and 47,XYY.

❏❏ **The phenotypic sex of an individual with 49,XXXXY is that of:**

A male.

❏❏ **Actual risks of chromosome aneuploidy in offspring of 47,XXX and 47,XYY individuals is:**

Not much higher than the general population.

❏❏ **In an individual with Turner syndrome, why is it important to rule out the presence of a Y chromosome in the chromosome constitution?**

To eliminate the risk of gonadoblastoma in internal gonads.

❏❏ **In an individual with a constitutional chromosome complement which is 47,XX,+21, the risk for a pregnancy being affected with trisomy 21 is:**

50%.

❑❐ **The risk for a man affected with fragile X mental retardation having a son also affected with this condition is:**

0% (because the X chromosome would be inherited from the mother and the Y from the father).

❑❐ **The most rapid and accurate way to diagnose fragile X in a family is that of:**

Molecular analysis, including Southern blot and PCR.

❑❐ **For a couple who has one child with trisomy 21, what is the empirical recurrence risk which would be provided for a second child being born with a trisomy?**

1-2%.

❑❐ **The theoretical risks for liveborn offspring having Down syndrome when a parent has a balanced t(14;21) Robertsonian translocation would be:**

1/3.

❑❐ **The etiology for tetraploidy in a conception is?**

Triple fertilization or failure of cell fission following the first DNA replication.

❑❐ **When two or more rare gene disorders, which are known to be closely mapped, occur in the same individual, what is suspected.**

Presence of a contiguous gene syndrome.

❑❐ **A deletion which is not detectable by classic cytogenetics is considered to be:**

A microdeletion.

❑❐ **Another term for complete androgen insensitivity, an X-linked recessive trait is that of:**

Testicular familiarization.

❑❐ **The most common reason for performing prenatal diagnosis is that of:**

Advanced maternal age.

❑❐ **The maternal age cut-off used for pregnant patients to have prenatal diagnostic procedures for aneuploidy is:**

35.

❏❏ **Advanced paternal age is associated with an increased risk for what genetic phenomenon?**

New mutation.

❏❏ **In what screening procedure is maternal age, maternal weight, race, diabetes status, and gestational age critical for interpretation?**

Maternal marker hormone screening.

❏❏ **Genetic diagnosis before embryonic implantation is considered to be:**

Preimplantation diagnosis.

❏❏ **A condition characterized by c̲oloboma, h̲eart defect, a̲nal atresia, r̲etardation, g̲enital abnormalities, and e̲ar anomalies is considered to be:**

CHARGE association.

❏❏ **The phenotype of an individual with two X chromosomes with a translocation of the sry+ gene segment to one X chromosome is that of:**

Klinefelter syndrome.

❏❏ **A product of conception in which both haploid chromosome complements are both parentally derived, because of the replication of a haploid sperm is termed:**

A complete mole.

❏❏ **The chromosome complement of a "partial mole" is generally that of:**

Triploidy.

❏❏ **The prenatal aneuploid screen which incorporates fluorescent *in situ* hybridization (FISH) will typically detect which aneuploid states:**

Trisomy 13, 18, 21, and sex-chromosome aneuploid states.

❏❏ **"Standard" genetic amniocentesis is performed at what gestational age?**

14-18 weeks.

❏❏ **"Early" genetic amniocentesis is performed at what gestational age?**

10-13 weeks.

❏❏ **A prenatal diagnostic procedure which has been reportedly associated with limb reduction defects, when performed before 10 weeks gestational age is that of:**

chorionic villi sampling (CVS).

❏❏ **Fetal blood sampling, also called cordocenteses, is associated with a fetal loss rate of:**

2-5%.

❏❏ **The generally accepted turn-around-time for genetic amniocentesis in receiving cytogenetic results is approximately:**

10 - 14 days.

❏❏ **Tetrasomy of the short arm of chromosome 12 is associated with what syndrome?**

Pallister-Killian syndrome.

❏❏ **In general, *de novo* apparently balanced chromosome rearrangements are associated with a risk for phenotypic abnormality to be:**

6-7% over and above the background risk.

❏❏ **You have a patient with an apparently balanced reciprocal translocation and she asks about the risks for a chromosomally unbalanced conception. You would inform her that the theoretical risk for having a chromosomally unbalanced pregnancy, with associated abnormalities, would be on the order of:**

50%.

❏❏ **The clinical practice arm of medical genetics which oversees certification of medical genetic personnel and oversees policies and approaches to genetic practices is that of:**

American College of Medical Genetics (ACMG).

❏❏ **Male germ cells which arise directly from the development of spermatogonia and through further division give rise to secondary spermatocytes are called:**

Primary spermatocytes.

❏❏ **The female germ cell precursors which arise directly from the nuclear division of primary oocytes through miosis I are:**

Secondary oocytes.

❏❏ **Male germ cell precursors are:**

Spermatogonia.

❏❏ **A type of in vitro fertilization procedure in which the sperm is injected into the cytoplasm of the egg is termed:**

Intracytoplasmic sperm injection (ICSI).

❑❑ **The gene that codes for the testis determining factor (TDF) is called:**

Sex determining region of the Y (SRY).

❑❑ **The uniform non-cellular layer which surrounds an oocyte is:**

Zona pellucida.

❑❑ **The genetic constitution or composition of an individual is referred to as the:**

Genotype.

❑❑ **The observed result of interaction of the genotype with environmental factors is the:**

Phenotype.

❑❑ **The mode of inheritance in which a gene is expressed in the heterozygous state is that of:**

A dominant.

❑❑ **Variable expressivity is most frequently associated with what type of inheritance pattern?**

Autosomal dominant traits.

❑❑ **The mating between close relatives is termed:**

Consanguinity.

❑❑ **Males are hemizygous for genes on what chromosome?**

X chromosome.

❑❑ **Vitamin D resistant rickets typically follows what mode of inheritance?**

X-linked dominant.

❑❑ **One of the most common and best characterized mitochondrial diseases is:**

Leber's hereditary optic neuropathy.

❑❑ **A change in the genetic material is referred to as:**

A mutation.

❑❑ **Those traits or diseases which are caused by the impact of many different genes is termed:**

Polygenic.

❑❑ **Cooley's anemia is also the name for:**

Thalassemia major.

❑❑ **A condition which is characterized by black urine and degenerative arthritis of the spine and large joints is most likely:**

Alkaptonuria.

❑❑ **Hemophilia A is an X-linked recessive condition which is due to what deficiency?**

Factor VIII.

❑❑ **Factor IX deficiency results in what disease?**

Hemophilia B.

❑❑ **Most inborn errors of metabolism which represent mutations in the genes for specific enzymes are inherited in what fashion?**

Autosomal recessive.

❑❑ **Blueish sclera and frequent bone fractures are suggestive of what condition?**

Osteogenesis Imperfecta.

❑❑ **A condition which is characterized by tissue elasticity and fragility, joint laxity and thinness of the skin with abnormal scaring, most likely represents one of the forms of:**

Ehlers Danlos.

❑❑ **Techniques which have been developed for the diagnosis of genetic diseases in embryos before implantation is termed:**

Preconceptual diagnosis.

❑❑ **The approximate procedure-related risk for miscarriage to occur following a genetic amniocentesis during the second trimester is approximately:**

1/500.

❑❑ **A twin pair in which both members exhibit the same phenotype or trait are considered to be:**

Concordant.

❑❑ **A twin pair in which one member exhibits a certain trait and the other does not is considered to be:**

Discordant.

❑❑ **An individual with one mutant allele on each of two different loci is considered to be a:**

Double heterozygote.

❑❑ **An autosomal dominant disease in which there is a deficiency of low density lipoprotein (LDL) receptors, resulting in elevated serum cholesterol and LDL-cholesterol is:**

Familial hypercholesterolemia.

❑❑ **The measure of fertility and therefore, contribution to the gene pool of succeeding generations is considered to be:**

Fitness.

❑❑ **A communication process, the objective of which is to provide individuals in families having a genetic disease or at risk for such a disease with information about their condition and to provide information that would allow couples at risk to make informed reproductive decisions is termed:**

Genetic counseling.

❑❑ **You consult with a couple, both of whom have autosomal recessive deafness, however, it has been determined that the cause of deafness in each is due to different alleles. You counseled that their risk for having offspring with congenital deafness would be:**

Close to 0 since all offspring would be heterozygotes for each allele but not homozygous for the same allele.

❑❑ **You consult with a couple, both of whom have autosomal recessive deafness, and it has been determined that the cause of the deafness is due to the same gene. You would counsel them that their risk for having an affected child would be on the order of:**

Approximately 100% since all offspring would be homozygous.

❑❑ **During fetal development, the failure of the neural folds to fuse at 20-30 days of embryonic development results in:**

Spina bifida.

❑❑ **The insult which results in anencephaly is generally considered to be multifactorial, possibly environmentally induced, sometimes related to a chromosome abnormality, but the primary lesion which results in anencephaly must occur at about what stage of fetal development?**

Within the first 3 weeks following conception.

❏❏ **Cleft palate is due to a fetal insult before what stage of pregnancy?**

8-10 weeks of development.

❏❏ **Approximately 50% of prenatally diagnosed omphalocele cases have an abnormal karyotype. The most common karyotypic abnormality as an etiology for an omphalocele include:**

Trisomy 13, Trisomy 18, triploidy.

❏❏ **An ultrasound diagnosis of oligohydramnios and fetal ascites is suggestive of:**

Renal agenesis or other renal problems.

❏❏ **The majority of fetuses with cystic hygroma have chromosome abnormalities. The most common abnormalities associated with fetal cystic hygroma are:**

45,X followed by a small number of infants with Trisomy 13, 18 and 21.

❏❏ **A 12-year-old girl presents to you with a recently diagnosed conductive hearing loss. Upon obtaining a family history you discover that her mother also wears a hearing aid and has a white forelock. What is a primary consideration in your differential diagnosis?**

Waardenburg syndrome.

❏❏ **In Triple Marker Screening, involving alpha-fetoprotein, unconjugated estriol, and human chorionic gonadotrophin, the pattern which is suggestive of trisomy 18 is:**

A low level of all three hormones.

❏❏ **An autosomal dominant trait which is expressed only in males but not females is considered to be:**

A sex-limited trait.

❏❏ **A condition which, when identified in the newborn period and the diet is restricted of dietary phenylalanine, will not develop microcephaly and profound mental retardation is:**

Phenylketonuria (PKU).

❏❏ **Your 25 year-old patient's father has just died of Huntington's disease. She requests genetic testing to determine whether she has the gene for Huntington's disease. You inform her that her first step would be:**

Seek genetic counseling to determine if knowing or not knowing if she has the gene for Huntington's disease is appropriate for her.

❏❏ **You have recently diagnosed a 5 year-old patient as having 45,X. What types of treatment are appropriate?**

Growth hormone and estrogen replacement.

❏❏ **The type of legal action which is considered when parents bring lawsuits against physicians alleging that they were not informed about the risk of having an abnormal child, is:**

Wrongful birth.

❏❏ **The type of legal action which might occur when an affected child brings lawsuits against their parents alleging that the parents knew the child would be abnormal, but continued the pregnancy, knowing that the child would have genetic abnormalities is:**

Wrongful life.

❏❏ **The occurrence of a sex-linked recessive condition in a female might be explained by:**

Lyonization or possibly 45,X.

❏❏ **The most serious health problem faced by children with Down syndrome is:**

Congenital heart disease.

❏❏ **A chromosomal syndrome which frequently involves Alzheimer's disease, if individuals survive beyond 35 years, is:**

Down syndrome.

❏❏ **An environmentally induced mimic of a genetic disorder is a:**

Phenocopy.

❏❏ **Patterns of anomalies which are derived from disturbances of a single developmental field, such as holoprosencephaly is considered to be:**

A field defect.

❏❏ **The time period following the formation of a primary spermatocyte from a stem cell to a mature sperm is approximately:**

60-70 days.

❏❏ **All of the primary oocytes have been formed by what developmental age?**

Three months prenatal development and the primary oolocytes remain suspended in late prophase of meiosis I until puberty.

❑❑ **Complex disorders which are associated with chromosome aberrations (usually microdeletions) affecting multiple unrelated genetic loci physically next to each other along the chromosome, are termed:**

Contiguous gene disorders.

❑❑ **A fifteen-year-old female patient presents with short stature, poorly developed secondary sex characteristics and has not begun menses. What is the likely diagnosis?**

Possible Turner syndrome.

❑❑ **A newborn infant has been diagnosed with Turner syndrome. What specialist should be consulted as soon as possible?**

Cardiologist, because of the high frequency of cardiac abnormalities in Turner infants.

❑❑ **The average age of onset for an individual with Huntington's disease is:**

35.

❑❑ **A dominant gene which is expressed in only some family members who have the gene is considered to show:**

Reduced penetrance.

❑❑ **A couple, both of normal physical stature, has one daughter and two sons with some form of dwarfism. What is the most likely mode of inheritance?**

Autosomal recessive.

CANCER GENETICS

❏❏ **Familial cancer is typically characterized by:**

Early age of onset in multiple family members.

❏❏ **In chronic myelogenous leukemia (CML), the proto-oncogene, abl, is activated by:**

Translocation involving chromosomes 9 and 22.

❏❏ **Generally, familial cancer syndromes follow what pattern of inheritance?**

Autosomal dominant with variable expression.

❏❏ **Tumor suppressor genes quite often are characterized by:**

Loss of heterozygosity.

❏❏ **A gene which normally encodes a protein which stimulates cell growth and division is known as:**

A proto-oncogene which, upon mutation, can become an oncogene that can lead to unregulated cell growth and division.

❏❏ **In cancer, which laboratory test frequently has diagnostic significance, prognostic implications, and can provide therapeutic direction?**

Cytogenetic findings.

❏❏ **Cancer cannot occur in which type of cells?**

In cells which are no longer capable of dividing.

❏❏ **One example of a tumor suppressor gene which also has properties of a proto-oncogene is that of:**

p53 gene.

❏❏ **Most cancers in humans are examples of:**

Somatic mutation.

❑❑ **The most important factor in enabling follow-up for patients with leukemia, myelodysplastic syndrome and other cancers is that of having an initial diagnostic laboratory study involving:**

Cytogenetics.

❑❑ **The most common cytogenetic finding in Fanconi anemia is:**

Increased chromosome breakage.

❑❑ **In childhood acute lymphocytic leukemia, what is the most favorable prognostic cytogenetic finding?**

Hyperdiploidy.

❑❑ **In childhood acute lymphocytic leukemia, what are some of the more unfavorable cytogenetic findings?**

t(4;11), hypodiploidy, balanced t(1;19).

❑❑ **What is one of the more frequently observed chromosomal abnormalities in pediatric T-cell anaplastic large cell lymphoma?**

t(2;5) - NPM/ALK fusion.

❑❑ **In adult acute nonlymphocytic leukemia, the most common chromosomal finding in the M2 stage is that of:**

t(8;21).

❑❑ **In adult acute nonlymphocytic leukemia, stage M3, the most common chromosomal finding is that of:**

t(15;17).

❑❑ **The inv(16) or del(16) is most commonly representative of:**

AML-M4.

❑❑ **The most common molecular cytogenetic approach for monitoring patients following opposite sex transplant is that of:**

Fluorescence *in situ* hybridization (FISH) involving centromeric probes for X and Y chromosomes.

❑❑ **Ewing sarcoma is most often characterized by a translocation involving which chromosomes:**

11 and 22.

❑❑ **The reasons for performing cytogenetics for myelo and lymphoproliferative diseases at presentation are:**

for diagnostic confirmation, for allowing eventual monitoring of therapy effectiveness and to monitor disease progression with future studies.

❑❑ **Congenital leukemias are generally caused by:**

Prenatal exposures.

❑❑ **The most common chromosome rearrangement in non-Hodgkin's lymphoma is:**

t(14;18).

❑❑ **Mantle cell lymphoma is most often characterized by which cytogenetic rearrangement?**

t(11;14).

❑❑ **Common chromosome rearrangements associated with secondary leukemia and/or MDS is that of:**

del(5q) and –7.

❑❑ **The preferred tissue for cytogenetic evaluation in the case of leukemias is that of:**

Bone marrow.

❑❑ **The preferred tissue for cytogenetics in the case of non-Hodgkin's lymphoma is:**

Lymph node or other involved tissue.

❑❑ **Approximately what percent of breast cancers are considered to be familial and associated with brca1 brca2?**

5%.

❑❑ **Leukemia occurs in what percent of patients with Down syndrome?**

1% (20 times the frequency in individuals without Down syndrome).

❑❑ **Neoplasms which are associated with deletions of chromosome 13q14 include that of:**

Retinoblastoma.

❑❑ **A fusion of bcr/abl is characteristic of:**

Chronic myelogenous leukemia (CML), or less likely, acute lymphocytic leukemia (ALL).

❑❑ **The t(8;14) frequently occurs in what type of lymphoma?**

Burkitt lymphoma.

❑❑ **Chronic myelogenous leukemia (CML) is most commonly associated with what chromosome rearrangement?**

t(9;22), also called the Philadelphia chromosome.

❑❑ **Normal genes which are altered or inappropriately expressed or over-expressed which can lead to neoplasia are referred to as:**

Oncogenes.

❑❑ **Normal genes which function to prevent the development of tumors are termed:**

Tumor suppressor genes.

❑❑ **The development of tumors such as retinoblastoma require two separate mutations. This theory is referred to as:**

Knudson's "two-hit" hypothesis.

❑❑ **The best possibility for "cure" in chronic myelogenous leukemia (CML) is:**

Allogenic bone marrow transplantation.

❑❑ **You are consulting with your newly diagnosed young man with cancer. You have explained that the proposed therapy will yield a high chance for cure but that he will probably have therapy induced infertility. One further option which you should offer is:**

Semen analysis and cryopreservation before initiation of therapy.

❑❑ **Anaplastic large cell lymphoma occurs most frequently in what age group?**

Pediatric population.

❑❑ **Trisomy 21 individuals have the highest predisposition to what type of cancer?**

Leukemia.

❑❑ **Bone marrow clonal chromosome abnormalities indicate:**

A myeloid or a lymphoid malignancy or another metastatic tumor.

❑❑ **The cytogenetic characteristics of patients with Fanconi anemia are:**

Increased spontaneous and induced chromosome breakage in lymphocytes and fibroblasts.

❑❑ **Deletion of 5q in bone marrow is most likely indicative of:**

Myelodysplastic syndrome (MDS).

❑❑ **Familial early age onset of cancer is indicative of:**

Genetic predispositional factors in cancer.

❑❑ **Chromosome analysis of bone marrow from individuals with myelodysplastic syndrome is important in order to:**

Determine base-line cytogenetics, identify prognostic indicators, monitor treatment efficiency, and to monitor disease progression.

❑❑ **Retinoblastoma is the most often associated with the deletion of which chromosome?**

13q.

❑❑ **Burkitt's lymphoma is associated with a rearrangement of chromosome band:**

8q24.

❑❑ **Genes which cause suppression of tumor growth if there is only one normal copy is that of:**

A tumor suppression gene.

❑❑ **One of the most common acquired chromosome abnormalities in myelodysplastic syndrome is that of:**

del(5q).

❑❑ **The proto-oncogene, ABL, is activated in chronic myelogenous leukemia (CML) by a:**

A chromosome translocation involving chromosomes 9 and 22, ABL/BCR.

❑❑ **Monitoring of bone marrow engraftment following opposite-sex bone marrow transplantation can be most easily accomplished by:**

Fluorescence *in situ* hybridization (FISH) with X and Y probes.

❑❑ **What percent of patients with neoplasms have a strong predisposition which has been inherited as a simple Mendelian trait from parents?**

5%.

❑❑ **Syndromes which have been found to be associated with multiple benign or multiple malignant neoplasms are of what Mendelian nature?**

Autosomal dominant with variable expression.

❑❑ **Chromosome abnormalities which are observed in cancer cells are typically considered to be constitutional or acquired abnormalities?**

Acquired abnormalities.

❑❑ **The malignancy which results in the MYC oncogene and an immunoglobulin locus generally results in:**

Burkitt lymphoma.

❑❑ **Increased chromosomal breakage, autosomal recessive inheritance, an increased frequency in the Jewish population and an increased risk of malignancy suggests the most likely diagnosis of:**

Bloom syndrome.

❑❑ **Bloom syndrome, Fanconi Anemia, and Ataxia-telangiectasia are autosomal recessive diseases which predispose to various types of cancer. These are considered to be:**

Chromosome instability or chromosome breakage syndromes.

❑❑ **In neoplasms, chromosomal deletions result in what event which leads to malignancy?**

Loss of heterozygosity (LOH).

❑❑ **Genes which, unless deleted, prevent formation of cancer are termed:**

Tumor suppressor genes.

❑❑ **Breast cancer is characterized by amplification of what genes?**

her2nu.

❑❑ **Specific genes which undergo mutation to cause cancer are termed:**

Oncogenes.

❑❑ **Burkitt's lymphoma results because of the activation of what oncogene?**

Myc.

❑❑ **CML results from the activation of what oncogene?**

Abl.

❑❑ **Non-Hodgkin's Lymphoma, particularly follicular lymphoma, most often is associated with the activation of what oncogene?**

bcl-2.

❑❑ **A chromosome translocation involving 8q24 and 14q32 is most commonly typical of:**

Burkitt's lymphoma.

❑❑ **Chromosome expressions which represent gene amplification in the karyotype are termed:**

Double minutes (DMs) or homogeneous staining regions (HSRs).

❑❑ **N-myc oncogene amplification is associated with what disease?**

Neuroblastoma.

❑❑ **A constitutional or acquired deletion of chromosome 13 is a predisposition to:**

Retinoblastoma.

❑❑ **The two most common genes which predispose to familial breast cancer are:**

brca1 and brca2.

❑❑ **A biochemical test which is widely used for both diagnosis and monitoring of cancer uses the levels of:**

Carcinoembryonic antigens (CEA).

❑❑ **The chromosomal translocation which is characteristic of synovial sarcoma is:**

t(X;18).

❑❑ **Alveolar rhabdomyosarcoma is characterized by what chromosomal translocation:**

t(2;13).

❑❑ **An individual with a bcr/abl rearrangement at the minor breakpoint has what disease?**

Acute lymphocytic leukemia (ALL).

❑❑ **The hypothesis which proposes that two events are necessary for malignancy occur, which was first described for the etiology for retinoblastoma and involves the presence of a hereditary first step followed by an environmental second step before the initiation of retinoblastoma to occur is termed:**

Knudson hypothesis.

❑❑ **Programed cell death is also referred to as:**

Apoptosis.

❑❑ **In the accelerated or blast phase of CML, the most common chromosome indicators of this disease acceleration are:**

A second Philadelphia chromosome, trisomy 8, trisomy 19, and isochromosome 17q.

❑❑ **In acute promyelocytic leukemia (M3), characterized by a t(15;17), the recommended mode of therapy is that of:**

Al-transretinoic acid (ATRA). Treatment with traditional modes of therapy for other subtypes of AML are contraindicated because of the induction of medical complications such as a potentially fatal pulmonary capillary leakage and frequently death.

❑❑ **Rearrangement of 11q23 is associated with multiple case types of mixed linage leukemia or myeloid lymphoid leukemia and disrupts what gene?**

MLL cluster.

❑❑ **Deletions of the long arm of chromosome 20 [del(20q)] are generally most often seen in what type of malignancies.**

Myeloid malignancies including polycycemia vera, MDS, and occasionally in AML.

❑❑ **The most frequently observed trisomy in hematopoietic malignancies is that of:**

Trisomy 8.

❑❑ **A common secondary change in the M-2 (AML) which is characterized by a t(8;21) is that of:**

Loss of the Y chromosome.

❑❑ **The loss of which chromosome is a common occurrence in elderly males and could also be seen in certain hematopoietic malignancies?**

Y chromosome.

CLINICAL CYTOGENETICS

☐☐ **The normal number of chromosomes in a human karyotype is:**

46 (22 pairs of autosome and 2 sex chromosomes).

☐☐ **The short arms of a chromosome are termed:**

p - for "petite".

☐☐ **The long arms of a chromosome are termed:**

q.

☐☐ **A chromosome nomenclature, 45,XX,t(14;21)(q11;q11), indicates:**

A female karyotype with 45 chromosomes which includes a balanced chromosome translocation, a carrier with an increased risk for having offspring with Down syndrome.

☐☐ **Three copies of all chromosomes is termed:**

Triploidy.

☐☐ **A single copy of one autosome is termed:**

Monosomy.

☐☐ **Three copies of a particular autosome is termed:**

Trisomy.

☐☐ **A partial monosomy of one autosome is termed:**

Autosomal deletion.

☐☐ **Partial trisomy of part of an autosome is termed:**

Duplication.

☐☐ **Four copies of all chromosomes in each cell is termed:**

Tetraploidy.

☐☐ **The incidence of Down syndrome at birth is:**

1 out of 680 liveborn infants.

❑❑ **A chromosome which has nearly equally sized short and long arms is termed:**

Metacentric.

❑❑ **A chromosome in which the centromere is closer to one end than the other is termed:**

Submetacentric.

❑❑ **A chromosome in which the centromere is near the end of the chromosome, with satellites on the short end is termed:**

Acrocentric.

❑❑ **Ultrasonic observations which include nuchal thickening or cystic hygroma, duodenal stenosis or atresia, and short femur length are frequently associated with:**

Down syndrome.

❑❑ **Advanced maternal age is generally considered to be a term used when:**

Mother is 35 years of age or older at the time of expected delivery.

❑❑ **When there is a presence of two or more cell lines with different karyotypes in a particular patient, the term used is:**

Mosaicism.

❑❑ **When one of the chromosome arms is duplicated and all of the material from the other arm is lost, so that the arm on one side of the centromere is a mirror image of the other, is termed:**

Isochromosome.

❑❑ **A clinically normal parent, who has an isochromosome for an acrocentric autosome such as i(21q), has a risk for a trisomic offspring of:**

100% - because only disomic or nullisomic gametes can be produced with the nullisomic gametes resulting in lethal fetal conditions.

❑❑ **When a break occurs on both sides of the centromere with the chromosomal material being inverted, this is correctly termed as a:**

Pericentric inversion.

❑❑ **When breaks and rearrangements occur on the same side of the centromere with the chromosome material between the breaks being flipped and rejoined, this is referred to as a:**

Paracentric inversion.

❑❑ **A child who exhibits growth retardation, telangiectasias on the face, and has an increased frequency of chromosomal breakage, the most likely diagnosis is:**

Bloom syndrome.

❑❑ **Chromosomal breakage syndromes, including Fanconi anemia, Bloom syndrome, and ataxia telangectasia have an increased risk for:**

Neoplasm.

❑❑ **Given the stages of a mitotic cell cycle being G1, S, G2, and M, at which stage is cytogenetic analysis performed.**

M (mitotic stage).

❑❑ **The single most common karyotype found in products of conception following first trimester miscarriage is:**

45,X, followed by trisomy 16, other autosomal trisomies and triploidy.

❑❑ **A chromosome nomenclature of 69,XXY indicates what diagnosis?**

Triploidy.

❑❑ **The tissue type which allows the most rapid cytogenetic evaluation of an individual is:**

Bone marrow.

❑❑ **In a routine evaluation of chromosomes, at which stage of mitosis are chromosomes generally analyzed?**

Metaphase.

❑❑ **In a high resolution chromosome study, at what stage of mitosis are chromosomes typically analyzed?**

Prometaphase (early metaphase).

❑❑ **In what year was the correct number of chromosomes in the human cell established?**

1956.

❑❑ **Women who are of advanced maternal age are at an increased risk to have a pregnancy affected with:**

Trisomy 21 and also other trisomies.

❏❏ **The most frequent mechanism for the cause of Down syndrome in children of couples over the age of 35 is:**

Maternal meiotic nondisjunction.

❏❏ **The most common autosomal chromosome disorder is:**

Trisomy 21.

❏❏ **A human triploid cell will have a chromosome count of:**

69.

❏❏ **Prenatal diagnosis can be accomplished by performing which procedures?**

Amniocentesis, chorionic villi sampling, and percutaneous umbilical cord sampling.

❏❏ **When both homologues are inherited from the same parent, this is termed:**

Uniparental disomy.

❏❏ **Genes which are in close proximity on the same chromosome are termed:**

Linked.

❏❏ **The 47,XYY syndrome is characterized by taller than normal stature and cystic acne. This syndrome is caused by what meiotic genetic mechanism?**

Paternal nondisjunction.

❏❏ **Triploidy is a frequent cause of miscarriage but is rarely seen in term liveborn infants. What are the mechanisms by which triploidy arises?**

Double fertilization or retention of a polar body.

❏❏ **A couple had a stillborn daughter with microcephaly. Subsequently, they had three spontaneous miscarriages and the fifth pregnancy resulted in a normal daughter. They recently had a son who was diagnosed with "Cri du Chat syndrome". What is the most likely cause of miscarriages?**

Malsegregation of a balanced reciprocal translocation from one parent.

❏❏ **Horizontal division of the centromere rather then longitudinal division of the centromere results in:**

Isochromosome.

❏❏ **Trisomy of chromosomes other than 13, 18, 21, and sex chromosomes results in:**

Miscarriage.

❏❏ **Chromosomal nondisjunction results in:**

Trisomy or monosomy.

❏❏ **X-chromosome inactivation occurs at what stage of fetal development?**

During early embryogenesis in female somatic cells but not in germ cells.

❏❏ **The total number of chromosomes present in a balanced carrier of a reciprocal translocation is:**

46.

❏❏ **Normal meoisis followed by an abnormality in mitosis during early embryonic development generally results in:**

Mosaicism.

❏❏ **A male with tall stature, behavioral disorders, acne, but fertile is characteristic of:**

47,XYY.

❏❏ **Severe central nervous system malformations, severe growth and mental retardation, polydactyly, cleft lip and palate are characteristic of:**

Trisomy 13.

❏❏ **Short stature, webbed neck, neonatal lymphodema and primary amenorrhea are characteristic of:**

Turner syndrome.

❏❏ **Mental retardation, failure to thrive, micrognathia, low-set and malformed ears and rocker bottom feet are characteristic of:**

Trisomy 18.

❏❏ **Mental retardation, epicanthal folds, Brushfield spots, low nasal bridge, protruding tongue and short broad hands are characteristic of:**

Trisomy 21.

❏❏ **Small testicles, infertility, and poorly developed secondary sex characteristics are characteristic of:**

Klinefelter syndrome.

❏❏ **The majority of first trimester abortions occur because of:**

An aneuploid state in the fetus.

❏❏ **Which of the human chromosomes are acrocentric?**

13, 14, 15, 21, and 22.

❏❏ **Congenital malformations, altered facial features, developmental delays and growth retardation are most likely representative of:**

Chromosome abnormalities.

❏❏ **The term for a missing portion of a chromosome is:**

Deletion.

❏❏ **The synonymous term for centric fusion is:**

Robertsonian translocation.

❏❏ **The single most common reason for prenatal diagnosis is that of:**

Advanced maternal age.

❏❏ **The most appropriate genetic test for an infant born with congenital heart disease which includes a left heart lesion is:**

FISH (fluorescence *in situ* hybridization) for 22q and chromosome studies.

❏❏ **The recurrence risk for a couple who has a previous child with trisomy 21 is:**

1-2%.

❏❏ **A chromosome without a centromere is termed?**

Acentric.

❏❏ **A deletion of a portion of the short arm of chromosome 5 [del(5p)] results in what syndrome?**

"Cri-du-chat".

❏❏ **An individual composed of a mixture of genetically different cells is called:**

Chimera (i.e., 46,XX/46,XY) or mosaic (i.e., 47,XX,+21/46,XX).

❏❏ **A chromosome with two centromeres is termed?**

Dicentric.

❏❏ **The cell type most commonly cultured from solid tissue for cytogenetic study is the:**

Fibroblast.

❑❑ **The condition in which there is only one homolog present for all chromosomes is:**

Haploid.

❑❑ **When chromosome segments are "flipped" 180° with the result being a reverse of gene sequences, this is termed:**

An inversion.

❑❑ **When only one chromosome 3 is present in a mitotic cell, this is termed:**

Monosomy 3.

❑❑ **When three copies of chromosome 13 are present in mitotic cells, this is termed:**

Trisomy 13.

❑❑ **The presence of multiple copies of each chromosome in a cell is termed:**

Polyploidy.

❑❑ **Detection of some specific microdeletions, prenatal diagnosis for aneuploidy, confirmation of loss of chromatin material identified by traditional banding and identification of marker chromosomes is best accomplished by what procedure?**

Fluorescence *in situ* hybridization (FISH).

❑❑ **Fanconi Anemia is characterized by what cytogenetic features?**

Increased spontaneous and induced chromosome breakage.

❑❑ **You have a Down syndrome infant in your practice who has a Robertsonian translocation. How many chromosomes does this patient have?**

46.

❑❑ **Centric fusion (Robertsonian translocations) involves which chromosomes?**

Acrocentric chromosomes (chromosomes 13, 14, 15, 21, 22).

❑❑ **At what stage of mitosis does spindle fiber formation occur?**

Metaphase.

❑❑ **X chromosome inactivation occurs during what stage of embryo development in the female?**

Early embryogenesis, approximately the 64-128 cell stage.

❑❑ **Chromosomes are most condensed during which stage of mitosis?**

Metaphase.

❑❑ **Sister-chromatid exchange is defined as the:**

Exchange of DNA between identical chromatids, without a loss or gain of genetic material.

❑❑ **Advanced maternal age is associated with:**

An increased risk of meiotic nondisjunction.

❑❑ **Horizontal division of chromosomes at the centromere results in:**

An isochromosome.

❑❑ **If one of the X chromosomes is inactivated in normal females, why do females with 45,X have abnormalities?**

Portions of the inactive X chromosome in normal females remain active.

❑❑ **A normal, non-disease producing genetic trait seen very rarely in the population is called a:**

Genetic polymorphism.

❑❑ **Prader-Willi, Miller Dieker, DiGeorge and Williams syndromes are examples of:**

Chromosome microdeletion syndromes.

❑❑ **Of the following chromosomal syndromes (Down, Prader-Willi, Klinefelter, XXX, and triploidy), which has the shortest life expectancy?**

Triploidy.

❑❑ **The chromosome staining techniques which best identifies satellite regions of acrocentrics is:**

NOR staining.

❑❑ **Your patient has a balanced Robertsonian translocation. How many chromosomes would be apparent on a karyotype?**

45.

❑❑ **What is the most common banding technique utilized in the United States?**

G-banding.

❑❑ **Chromosome abnormalities which are present at birth and are most often present at conception are referred to as:**

Constitutional chromosome abnormalities.

❑❑ **The probability that a conception with trisomy 21 will be liveborn is approximately:**

20%.

❑❑ **If a trisomy is found in a spontaneous abortion in your patient, you advise her that the possibility that the next pregnancy resulting in a trisomy is approximately:**

2%.

❑❑ **You have a couple in your practice who has had a child born with a partial deletion and partial duplication for the same chromosome. The most likely parental chromosome abnormality which would predispose to this condition would be:**

Paricentric inversion.

❑❑ **What is the most common mechanism for tetraploidy?**

Post-fertilization error with failure of fertilized egg to cleave after the first mitotic replication.

❑❑ **Paternal uniparental disomy of chromosome 11, results in:**

Macroglossia and overgrowth syndrome (Beckwith-Wiedemann syndrome).

❑❑ **Uniparental paternal disomy of chromosome 15 most likely results in:**

Mental retardation, seizures, absence of speech, and Angelman syndrome.

❑❑ **The molecular cytogenetic technique which involves hybridization of tumor DNA with normal DNA in order to detect gains and losses of genetic material in the genome is termed:**

Comparative genomic hybridization (CGH).

❑❑ **A constitutional or acquired deletion of chromosome 11p (11p13) is associated with:**

Aniridia / Wilms tumor complex.

❑❑ **Patients with one of the rare autosomal recessive breakage syndromes has an increased risk for:**

Neoplasms of various types.

❏❏ **The initials FISH, which represents a molecular cytogenetic technique represent:**

Fluorescent *in situ* hybridization (FISH).

❏❏ **The "p" which designates the short-arm of a chromosome stands for:**

Petite.

❏❏ **The two strands of a metaphase chromosome which are exact DNA replicas of each other are termed:**

Sister chromatids.

❏❏ **The term chromosome was derived from Greek words which mean:**

Colored body.

❏❏ **The two number 7 chromosomes seen in a normal human chromosome complement are termed:**

Homologues.

❏❏ **The two main chemical components of the chromosome are:**

DNA and protein(histones).

❏❏ **The terminal extremities of the short and long arms of chromosomes are termed:**

Telomeres.

❏❏ **A specialized region of DNA, which during mitosis provides the site at which the spindle apparatus can be anchored is termed:**

Centromere.

❏❏ **The incidence of chromosome abnormalities in recognized first trimester pregnancy losses is:**

At least 60%.

❏❏ **The frequency of chromosome abnormalities at livebirth is approximately:**

1 out of 150.

❏❏ **A chromosome banding technique which specifically stains the nucleolar or organizing region is:**

NOR (silver staining).

❏❏ **Meiotic nondysjunction will result in:**

Full chromosome aneuploidy.

❏❏ **Mitotic nondysjunction typically results in:**

Mosaicism.

❏❏ **A translocation which results in an equal exchange of chromosome material between two chromosomes is termed:**

Reciprocal translocation.

❏❏ **A fusion of the terminal segments of the short arm or long arm of a chromosome to another terminal segment of a short arm or a long arm of another chromosome is termed:**

Telomeric fusion.

❏❏ **The term "aneusomie de recombination" indicates:**

A resulting chromosome unbalanced state due to an inversion cross-over during meiosis.

❏❏ **How many chromosome breaks are required for a chromosome insertion to occur?**

Three.

❏❏ **Breakage in both the short and long arm of a single chromosome, followed by joining of these two ends, results in a:**

Ring chromosome.

❏❏ **Ring chromosome formation results in a deletion of:**

The short and long arm material distal to the breaks.

❏❏ **The region of the Y chromosome which is homologous to sequences on the X chromosome is termed:**

Pseudoautosomal region.

❏❏ **If a person has four X-chromatin bodies (Barr bodies) in somatic cells, how many X chromosomes are present?**

Five.

❏❏ **Homologous chromosomes which vary in centromeric size (C-band size) are considered to be:**

Normal chromosome variants.

❑❑ **Four copies of each chromosome in a cell represents:**

Tetraploidy.

❑❑ **The syndrome which is classically associated with high levels of sister chromatid exchanges (SCEs) is that of:**

Bloom syndrome.

❑❑ **A "mirror-image" chromosome which reflects an exact duplication of the short or long arm of a particular chromosome is termed:**

Isochromosome.

❑❑ **An inactivated X chromosome which upon staining, is manifested as a darkly stained mass in somatic cells of mammalian females is termed:**

Barr body.

❑❑ **The cytologically identifiable point of crossover between members of homologous chromosomes during meiosis is termed:**

Chiasma.

❑❑ **An organism or tissue which is composed of two or more genetically distinct subpopulation of cells, derived from two different zygotes is termed:**

Chimera.

❑❑ **Chromosomes with a total lack of a short arm and whose centromere is located at one end with no short arm are termed:**

Telocentric chromosomes (which do not occur in humans).

❑❑ **A banding method which stains opposite of G-banding is termed:**

Reverse banding (R-bands).

❑❑ **A solution in which the extracellular media is less concentrated than the amount of solute that is intracellular and used during cell harvest for cytogenetic study is termed:**

Hypotonic solution.

❑❑ **The most common fixative used in chromosome harvest is that of:**

Methanol/Acetic acid (3:1).

❏❏ **The solution which inhibits the formation of spindle fibers during cell culture is:**

Colcemid.

❏❏ **The most common biological agent used in immortalization of lymphocyte cell cultures, i.e., lymphoblastoid cell lines, is that of:**

Epstein Bar virus (EBV).

❏❏ **The most common dye used to stain cells to determine cell viability is that of:**

Trypan-blue.

❏❏ **The staining method by which both dark and light differential staining is induced along the lengths of chromosomes is termed:**

Banding.

❏❏ **A cytogenetic phenomenon which results in intrauterine growth retardation, in which fetal chromosomes are normal, is that of:**

Confined placental mosaicism (CPM).

❏❏ **The syndrome which is characterized by craniofacial anomalies, mental retardation, and a deletion of the short arm of chromosome 5 is that of:**

Cri-du-Chat.

❏❏ **A structurally rearranged chromosome in which no part can be identified is referred to as a:**

Marker chromosome.

❏❏ **A chromosomal fragment which does not contain a centromere is termed:**

Acentric fragment.

❏❏ **An agent which causes chromosome breakage is called:**

Clastogen.

❏❏ **A molecular cytogenetic assay which screens for DNA copy number changes within a complete genomic complement and maps these changes to normal chromosomes is called:**

Comparative genomic hybridization (CGH).

❏❏ **The area of clinical genetics which is concerned with the diagnosis and management of congenital anatomic abnormalities is:**

Dysmorphology.

❑❑ **An abnormal form, shape, or position of part of the body caused by mechanical forces is termed a:**

Deformation.

❑❑ **A pattern of multiple anomalies which is derived from a single prior anomaly or a mechanical factor, is termed a:**

Sequence.

❑❑ **A pattern of multiple anomalies thought to be pathologically related but not known to represent a single simple sequence is termed a:**

Syndrome.

❑❑ **Most of the constitutive heterochromatin resides on what part of the chromosome?**

Around the centromere.

❑❑ **Molecular cytogenetic methods which result in the differential fluorescent staining of each chromosome pair in a different color are:**

Multi-colored FISH (M-FISH) and Spectral Karyotyping (SKY).

❑❑ **The cell type which is cultured from skin or other tissue is typically?**

Fibroblast.

❑❑ **Cytogenetic fragile X expression in culture is initiated by:**

Folic acid deficient media or the use of a folic acid antagonist.

❑❑ **The banding method which stains opposite, from G-bands is:**

R-banding (reverse banding).

❑❑ **The chemical utilized in cytogenetics which prevents the formation of spindle fibers is:**

Colcemid.

❑❑ **Complete monosomy of all of the chromosomes are lethal with the exception of:**

The X chromosome.

❐❐ **According to the Lyon hypothesis, how many X chromosomes would be inactivated in a person with 48,XXXY?**

Two.

❐❐ **The arrangement of chromosomes in pairs in order of size and centromere location is called a:**

Karyotype.

MOLECULAR AND BIOCHEMICAL GENETICS

❏❏ **A molecular diagnostic method which includes gel electrophoresis with the use of specific probes, is called:**

Southern blot analysis.

❏❏ **A method which involves very selective and rapid amplification of target DNA or RNA sequences is termed:**

PCR (polymerase chain reaction).

❏❏ **Problems involved in gene therapy include:**

Stable incorporation of the gene, targeting of the gene to the appropriate tissue, synthesizing large quantities of the gene to be transplanted and determining the appropriate vector.

❏❏ **The molecular cytogenetic technique fluorescence *in situ* hybridization (FISH), is currently being used for:**

The detection of certain microdeletions, prenatal diagnosis for aneuploidy, confirmation of cytogenetic uncertainties exposed by traditional banding, and identification of marker chromosomes.

❏❏ **In order to perform linkage studies for presymptomatic testing for a late onset autosomal dominant disease, DNA samples are needed from which family members?**

Both affected and unaffected.

❏❏ **Myotonic dystrophy is an autosomal dominant disorder with many different features and it is a condition which frequently gets progressively worse in succeeding generations. The explanation for this observation is:**

Expansion of trinucleotide repeat units (called anticipation).

❏❏ **Polymerase chain reaction (PCR) is used for:**

Amplification of DNA molecules.

❑❑ **One of the most common genetic causes of mental retardation in the human population is that of:**

Fragile X syndrome.

❑❑ **Traits which require the interaction of multiple genes for their expression are termed:**

Polygenic.

❑❑ **If two genes undergo recombination 10% of the time, the gene distance between the two loci is:**

10 centimorgans.

❑❑ **The determination of the molecular basis of human diseases permits:**

The identification of carriers, prenatal diagnosis, genotype/phenotype correlations, and the development of therapeutic strategies.

❑❑ **The chromosome banding technique which will best identify satellite regions of acrocentric chromosomes is that of:**

Nuclear organizing region staining (NOR) - Silver staining.

❑❑ **The presence of two chromosomes of a pair, both inherited from one parent, with no representative of that chromosome from the other parent is the definition of:**

Uniparental disomy.

❑❑ **The highly repetitive DNA sequences found near the centromeres of chromosomes is referred to as:**

Satellite DNA.

❑❑ **Osteogenesis Imperfecta is due to mutations which result in defective:**

Collagen.

❑❑ **The most common cystic fibrosis mutation is:**

Delta 508 (Δ508).

❑❑ **The chromosome banding technique which produces a pattern nearly identical to G-banding techniques is:**

Q-banding.

❑❑ **Considering clinical implications, euchromatin is generally:**

Transcriptionally active.

❑❑ **In considering clinical implications of heterochromatin:**

It is generally transcriptionally inactive and not clinically relevant.

❑❑ **Nondisjunction during meiosis generally causes:**

Aneuploid gametes.

❑❑ **Factors which may result in DNA diagnostic tests being unreliable in clinical practice include:**

Sample mix up, genetic recombination, nonpaternity and heterogeneity of the genetic locus which is responsible for the disease.

❑❑ **A boy with cystic fibrosis is homozygous for the delta 508 mutation. His mother is a carrier for the delta 508, but his father is not. How would this be explained?**

Maternal uniparental disomy, paternity (most likely) or less likely, a new mutation.

❑❑ **The cause of congenital adrenal hyperplasia is:**

21 hydroxylase deficiency.

❑❑ **Muelerian inhibiting substance (MIS) works by:**

Inhibiting growth of the paramesonephric tubules.

❑❑ **A primary consideration for diagnosis in a 3-year-old child with progressive developmental delay, joint stiffness, corneal clouding, hepatomegaly, and course facial features is that of:**

Hurler syndrome (MPS Type I).

❑❑ **Why do red blood cells from patients with Sickle Cell anemia sickle?**

Their hemoglobin aggregates at low oxygen tensions.

❑❑ **In animal trials where drugs show no effect on pregnancy, what can be concluded about the drug usage in humans during pregnancy?**

The drug may be safe but further data would be needed.

❑❑ **An elevated serum phenylalanine level in a neonate can occur due to:**

Deficient activity of phenylalanine hydroxylase, faulty biopterin formation and maternal PKU.

❑❑ **An infant with hyperammonemia, hypoglycemia, and metabolic acidosis may be found in an infant with:**

An inborn error of metabolism.

❑❑ **In female patients with phenylketonuria, when can the dietary restriction of phenylalanine be discontinued?**

After reproductive age.

❑❑ **In male patients with phenylketonuria, when can the dietary restriction of phenylalanine be discontinued?**

After 5 years of tolerance.

❑❑ **Your 5 month-old patient has been admitted because of a new onset of seizures. In the ER, the serum glycose was 28 mg/dl (Normals = 70-100). This child has been healthy since birth and this is the couples first child who has been breast fed since birth. Two days ago the mother started the infant on formula foods and jarred fruits. Among the long list of diagnoses you should consider, you should include:**

Galactosemia.

❑❑ **The majority of inborn errors of metabolism diseases follow what pattern of inheritance?**

Autosomal recessive.

❑❑ **The critical factor in consideration of including a screen in newborn screening panels is that of:**

Whether the condition can be treated and preventative measures can be undertaken.

❑❑ **Your female patient has recently been shown by PCR that she has inherited both chromosome 15 homologs from her father (and none from her mother). What is your patient's diagnosis?**

Angelman syndrome.

❑❑ **Reasons for DNA diagnostic errors in clinical practice include:**

Non-paternity, sample mix up, and possible gene recombination between a marker allele and the disease in question for indirect genetic tests.

❑❑ **You have an infant in your practice who is homozygous for the delta F508 mutation. His father is a carrier of the delta 508 mutation and his mother does not have the delta F508 mutation or any other detectable mutation. What is the most likely explanation?**

Paternal uniparental isodisomy for chromosome 7, or less likely, a new maternal mutation.

❏❏ **The phenomenon of anticipation is generally best explained by:**

Expansion of trinucleotide repeats.

❏❏ **The term which describes a variable phenotypic expression with the same genotype is that of:**

Pleiotropy.

❏❏ **The procedure which is utilized to identify specific microdeletions, prenatal aneuploidy, and identification of small chromosomes unidentified by cytogenetics is that of:**

Fluorescence *in situ* hybridization (FISH).

❏❏ **The observation of homogeneously staining regions and double minutes is an example of:**

Gene amplification, caused by drug resistance or disease acceleration.

❏❏ **Phenotypic traits which require the interaction of multiple genes in order to be expressed are considered to be:**

Polygenic.

❏❏ **If the genetic distance between two loci is 20 centimorgans, what is the recombination expectation?**

20%.

❏❏ **The substitution of one amino acid for another amino acid is considered to be that of what type of mutation.**

A missense mutation.

❏❏ **Both Duchenne and Becker Muscular Dystrophy have been mapped to the same locus on the X chromosome. They are therefore considered to be what type of disorder.**

Allelic disorders.

❏❏ **You have a patient in your office who asks about the reliability of testing for cystic fibrosis carrier state. She has no family history of cystic fibrosis but the carrier state detection frequency is approximately what percent?**

90% utilizing the laboratories which include the largest number of panels.

❏❏ **The molecular test to determine the level expression of messenger RNA from a specific gene is that of:**

Northern blot analysis.

❏❏ **The chromosome banding technique which specifically identifies the heterochromatic and centromeric areas of the chromosome are that of:**

C-bands.

❏❏ **In cystic fibrosis, the most common mutation is that of:**

Delta F508.

❏❏ **Considering chromosome material which is euchromatic or heterochromatic, which is the most clinically important?**

Euchromatic material.

❏❏ **What proportion of the total human genome actually encodes protein?**

Approximately 5%.

❏❏ **Meiotic nondisjunction results in:**

Aneuploidy at conception.

❏❏ **You have a young patient in your office who clinically is developmentally delayed, has osteoporosis and ectopic lenses. Laboratory urine studies show an increase in urinary homocystine. The most likely diagnosis is that of:**

Homocystinuria.

❏❏ **The most common result of 21 hydroxylase deficiency is that of:**

Congenital adrenal hyperplasia.

❏❏ **Among the following, which is the most likely to be characteristic of Hurler syndrome, tall stature, macroorchidism, webbing of the neck, brittle bones, coarsening of facial features?**

Coarsening of facial features.

❏❏ **Accumulation of lipids, mucopolysaccharides and glycoproteins, suggests a possible diagnosis of:**

Lyosomal storage diseases.

❏❏ **You have a couple in your practice who both have albinism and consult you regarding the risk for their children having the same condition. You review their biochemical status and determine that they both have the same genetic type of albinism. You counsel them that their risk for having an affected child is:**

100%.

❑❑ **You have a couple in your practice who both have albinism, but upon further investigation you determine that they have an absence of pigment due to different genetic alleles. Although both are autosomal recessive traits, these are not located on the same chromosome and are not alleles. You inform them that the risk for having affected offspring is:**

Close to zero.

❑❑ **The phenomenon of two different genes being located on the same chromosome is termed:**

Linkage.

❑❑ **When performing gene mapping studies, a high lod score indicates:**

Linkage of genes.

❑❑ **One agent which is considered to be a mutagen, a teratogen, and a carcinogen is:**

Radiation.

❑❑ **Which prenatal genetic diagnostic procedure is most susceptible to error because of maternal cell contamination?**

Chorionic villus sampling (CVS).

❑❑ **The objective for newborn genetic screening is:**

To detect genetic disease in which treatment is effective to reduce morbidity and mortality.

❑❑ **Advanced maternal age most significantly increases the risk for:**

Chromosome aneuploidy (trisomies, etc.).

❑❑ **The primary sex determining genes on the Y chromosome which induces testicular development is termed:**

The SRY gene complex.

❑❑ **The phenomenon of anticipation is most commonly associated with genetic disorders caused by:**

Triplet repeat expansions.

❑❑ **What is the gene frequency and carrier frequency of the Tay Sachs mutation in the Eastern European population?**

1/60 and 1/30, respectively.

❏❏ **Persons in families with affected members having myotonic dystrophy tend to be more severely affected with successive generations. This phenomenon is called:**

Anticipation.

❏❏ **A molecular biology method commonly utilizing amplification of small selected segments of nucleic acids is.**

Polymerase chain reaction (PCR).

❏❏ **The production of large numbers of identical copies of particular DNA fragments or the production of DNA identical organisms is termed:**

Cloning.

❏❏ **When DNA is cleaved with one or more restriction enzymes and run on an agarose gel where the fragments are separated by size is termed:**

Southern blotting.

❏❏ **The first genetic component involved in gene expression is:**

RNA.

❏❏ **RFLP is an abbreviation for:**

Restriction Fragment Linked Polymorphism.

❏❏ **A DNA index profile is informative about:**

Marked chromosome aneuploidy.

❏❏ **The double helix refers to:**

The molecular structure of DNA.

❏❏ **What are the four nitrogen containing basis which make up the structure of DNA?**

Adenine (A) and guanine (G) which are purines and cytosine (C) and thymine (T) which are pyrimidines.

❏❏ **Compared to DNA, the structure of RNA includes the substitution of uracil for:**

Cytosine.

❏❏ **The type of RNA which makes up part of the polyribosomes and is synthesized in the nucleolus is:**

Ribosomal RNA.

❏❏ **Crucial elements in the translation of genetic material into protein molecules are:**

Transfer RNA (tRNA).

❏❏ **The type of RNA which occupies the essential connecting link between information contained in a gene and its end result as the specific amino acid sequence of a protein is:**

Messenger RNA (mRNA).

❏❏ **How many different amino acids comprise proteins?**

20.

❏❏ **On the average, approximately how many crossovers (recombinations) occur during a meiotic division?**

Approximately 30-40 (or 1-2 per chromosome).

❏❏ **A specific position or location on a chromosome is referred to as a:**

Locus.

❏❏ **Alternate forms of a gene are termed:**

Alleles.

❏❏ **Genes which encode proteins which are known or thought to be involved in the disease process are termed:**

Candidate genes.

❏❏ **Stretches of DNA which are located within the gene, transcribed into RNA, but then spliced out before the RNA is translated into protein are termed:**

Introns.

❏❏ **Enzymes which can cut within a DNA strand and have the ability to cut double-stranded DNA at a specific nucleotide sequence are:**

Restriction enzymes.

❏❏ **An enzyme which allows single-stranded RNA to be converted to double-stranded DNA is:**

Reverse transcriptase.

❏❏ **A short single strand of DNA molecule which is typically 14-30 nucleotides long is called:**

An oligonucleotide.

❏❏ **A hybridization approach which is used to identify the specific messenger RNAs which are complimentary to the label probe is called:**

Northern blotting.

❏❏ **A mutation which results from a nucleotide change that alters amino acid encoded by a particular three-based codeon is termed:**

Missense mutation.

❏❏ **An insertion or deletion of a small number of nucleotides that are not a multiple of 3 results in what type of mutation?**

Frame-shift mutation.

❏❏ **The genetic term which describes the distance between loci is that of:**

Centimorgan (cM).

❏❏ **The statistical term that is generally calculated to evaluate the significance of linkage analysis is called a:**

Lod score (or Z). The Lod score compares the likelihood of obtaining the test data if the two loci are indeed linked.

❏❏ **The term which is generally used to describe the treatment of human disease by transfer of genetic material (DNA or RNA) into the patient is that of:**

Gene therapy.

❏❏ **Therapeutic gene modification which is introduced into all cells of the body is termed:**

Germ line gene therapy.

❏❏ **When a genetic modification is restricted exclusively to somatic cells with no effect on the germ line, this is termed:**

Somatic gene therapy.

❏❏ **The treatment of human disease by introduction of recombinant nucleic acid sequences into the cells of a patient is termed:**

Gene therapy.

❏❏ **An intermediate size expansion of a triplet repeat section (such as the CGG repeat in Fragile X syndrome) that does not itself cause the disease but is unstable when transmitted and is likely to increase in size during meiosis (in the mother in the case of Fragile X syndrome) is termed a:**

Premutation.

❏❏ **In molecular genetics, a labeled DNA or RNA sequence which is used to detect the presence of a complementary sequence by molecular hybridization is considered to be a:**

Probe.

❏❏ **A chemical or physical agent that produces or raises incidence of congenital malformations is a:**

Teratogen.

❏❏ **If a gene has been mapped but the specific gene has not been identified, the testing method to determine the presence of a particular gene is called:**

Linkage analysis.

❏❏ **The production of multiple copies of a DNA sequence is considered to be:**

Amplification.

❏❏ **The term which is used to refer to the inactivation of a specific target gene by homologous recombination methods is termed:**

Knockout.

❏❏ **A cloning vector propagated in yeast which can carry large inserts of DNA up to 1 megabase in length is termed:**

Yeast artifical chromosome (YAC).

❏❏ **Disorders which are caused by either quantitative or qualitative abnormalities of hemoglobin are termed:**

Hemoglobinopathies.

❏❏ **Which hemoglobin represents about 70% of the total hemoglobin at birth but less than 1% in adult life?**

Hemoglobin F (fetal hemoglobin).

❏❏ **An abnormal hemoglobin which aggregates and distorts red blood cells into a sickle shape is:**

Hemoglobin S.

❑❑ **You have a infant in your practice who is of African-American ancestry and presents to you with failure to thrive, recurrent infections, cardiac enlargement, hepatomegaly, abdominal pain, swelling of the hands and feet, and hematuria, a primary consideration in your differential diagnosis should include:**

Sickle Cell disease.

❑❑ **The hemoglobinopathy which is most prevalent in Southeast Asian population is that of:**

Alpha Thalassemia.

❑❑ **You have a young infant in your practice who was normal at birth but gradually develops gastrointestinal problems, psoriasis of the liver and cataracts. When milk is eliminated from his diet, his condition stabilizes. A primary diagnostic consideration should be:**

Galactosemia.

❑❑ **In the case of alpha-1-antitrypsin disease, ZZ homozygotes develop what medical complications early in life?**

Early onset lung disease, including destruction of pulmonary alveoli and chronic obstructive lung disease or emphysema.

❑❑ **A genetic condition which may lead to prolonged apnea of one to several hours after administration of succinylcholine, a muscle relaxant in anesthesia, is that of:**

Pseudocholinesterase deficiency.

❑❑ **A condition in which children are normal at birth but become mentally retarded when they ingest phenylalanine is:**

Phenylketonuria (PKU).

❑❑ **Sphingolipidoses, Glycoproteinoses, Mucolipidoses and Mucopolysaccharidoses are examples of what types of genetic diseases?**

Lysosomal storage diseases.

❑❑ **You have an eight month old infant in your practice whose parents are of Eastern European ancestry, and the child was normal at birth but has recently developed seizures, muscular weakness, apparent progressive motor and mental degeneration, and has a "cherry red spot" on the retina. A primary consideration in your differential diagnosis should be that of:**

Tay Sachs disease.

❏❏ **A condition which is characterized by no hexosaminidase A and a deficiency of hexosaminidase B is that of:**

Sandhoff disease, which is phenotypically similar to Tay Sachs disease but also may include more cardiac involvement and mild hepatosplenomegaly.

❏❏ **A condition in which there is a deficiency or absence of glucocerebrosidase which causes a build up of the substrate, glucosyl ceramide, is that of:**

Gaucher disease.

❏❏ **A disorder caused by a defective arylsulfatase A with an onset of approximately two years characterized by hypotonia, muscle weakness, unsteady gait, rigidity, seizures and mental retardation is that of:**

Metachromatic leukodystrophy.

❏❏ **Among the mucopolysaccharidoses (MPS), all of the eight subtypes are due to autosomal recessive inheritance except for:**

MPS type II (Hunter syndrome or Iduronidate sulfatase).

❏❏ **A condition which is characterized by decreased gastrointestinal absorption and transport of copper, which leads to decreased serum ceruloplasm with an onset of 6-8 weeks of life which is characterized by cerebral degeneration and sparse hair which breaks easily, is that of:**

Menkes disease.

❏❏ **You have a newborn in your practice who displays poor feeding, apathy, a high pitched cry, has developed seizures, and also has a characteristic urine odor of maple syrup most likely has:**

Maple Syrup Urine disease (MSUD), a form of branch chain keto acid decarboxylation deficiency in which the body cannot break down branch chain amino acids.

❏❏ **An X-linked recessive disorder caused by a complete deficiency of hypoxanthine guanine phosphoribosyltransferase (HPRT) which results in a phenotype characterized by a self-mutilating behavior, choreoathetosis, spasticity, and mental retardation, typically presenting at about 3-4 months of age is known as:**

Lesch-Nyhan syndrome.

❏❏ **The science which deals with immunology and genetic factors and issues is that of:**

Immunogenetics.

❑❑ **Immunoglobulins are mediators of the immune response in HLA related disease and with foreign antigens. The three main immunoglobulin gene families are:**

Heavy chain (H), Kappa light chain (K), Lambda light chain (λ).

❑❑ **The most severe of all immunodeficiency diseases in which both cellular and humoral immunity are profoundly affected which results in an affected individual having virtually no immune response is termed:**

Severe combined immunodeficiency syndrome (SCIDS).

❑❑ **Unique DNA sequences, which are protein coding sequences, present in only a single or a few copies, comprise approximately what percent of the total DNA in an individual?**

60%.

❑❑ **Highly repetitive DNA sequences which are untranscribed sequences present in tandem repeats at the heterochromatic regions of chromosomes makes up approximately what percent of the total genome?**

10%.

❑❑ **A molecular genetic technique which is used to obtain information on the size and amount of mutant protein in cell extracts from patients with genetic diseases is termed:**

Western blotting.

❑❑ **What is the term of the organism used when multiple copies of specific cloned genes are microinjected into fertilized mouse eggs followed by the search for their presence and expression in the mouse at embryonic stages or after birth?**

Transgenic mice.

❑❑ **RNA viruses which encode a reverse transcriptase so that they are transcribed into DNA upon entering the host-cell are termed:**

Retroviruses.

❑❑ **The antibiotic which has been known to be associated with enamel hypoplasia and caries of fetal teeth, as well as permanent discoloration of teeth when prenatal exposure occurs after 5-6 months gestation, is:**

Tetracycline.

POPULATION GENETICS

❑❑ **A specific physical location of a gene on a chromosome is termed:**

A locus.

❑❑ **When identical genes are present at the same locus on both homologues, this is termed:**

Homozygosity.

❑❑ **When there are two different alleles at the same locus on homologous chromosomes, this is termed:**

Heterozygosity.

❑❑ **When a dominant mutant gene is inherited, but not expressed, the term used is:**

Nonpenetrance.

❑❑ **When there are differences as to how or what degree a particular gene is expressed in an individual, this is known as:**

Variable expressivity.

❑❑ **When an environment causes a phenotype which produces a clinical picture similar to that produced by a specific genotype, this is termed:**

Phenocopy.

❑❑ **The law which is used to relate the frequencies of genotypes at a single Mendelian locus to the phenotype frequencies in a population is:**

The Hardy-Weinberg law.

❑❑ **Approximately what percent of all cases of achondroplasia result as new mutations?**

90%.

❑❑ **The term used to define the improvement of the human species by selective breeding is that of:**

Eugenics.

❑❑ **Assuming that the disease frequency for cystic fibrosis is 1/1600 in a certain population, what is the CF gene frequency, the normal gene frequency, the homozygous normal frequency, the homozygous abnormal frequency and the heterozygote (carrier) frequency?**

Using the Hardy-Weinberg law $p^2 + 2pq + q^2 = 1$; $q^2 = 1/1600$ (disease frequency) and $q = 1/40$ (gene frequency). The following frequencies can, therefore be calculated:

    p = 39/40 normal gene frequency
    q = 1/40 abnormal gene frequency
    p^2 = 1521/1600 (.95) homozygous normal
    q^2 = 1/1600 homozygous abnormal
    $2pq$ = 78/1600 (0.48) carriers (heterozygotes)

❑❑ **The science which involves the influence of genes on the determination of the response to drug therapy is termed:**

Pharmacogenetics.

❑❑ **Traits which are due to a combination of genetic and nongenetic factors are termed:**

Multifactorial.

❑❑ **The term used in medicine to define conditions in which only complex genetic factors are involved, is that of:**

Polygenic.

❑❑ **The term utilized to indicate a problem which is present at birth is that of:**

Congenital.

❑❑ **Patterns of anomalies, all of which are pathologically related, is termed:**

Syndromes.

❑❑ **An agent which can produce a permanent alteration of structure or function in an organism after exposure during embryonic or fetal life is termed:**

A teratogen.

❑❑ **An embryo is most sensitive to damage by a teratogen at what stage of gestation?**

Between two and ten weeks after conception (4-12 weeks following the beginning of the last menstrual period).

❑❑ **Teratogen exposure during the first two weeks after conception will result in:**

Either no effect or embryo lethality.

❐❏ **What percent of all congenital anomalies are thought to be a result of teratogenic factors?**

10%.

❐❏ **Chronic severe use of alcohol during pregnancies often results in:**

Fetal alcohol syndrome (FAS) or fetal alcohol effects (FAE).

❐❏ **The features of fetal alcohol syndrome (FAS) include:**

Growth deficiency, mental retardation, behavioral disturbances, hypoplastic midface, a long and flat philtrum as well as a narrow upper lip vermillion.

❐❏ **An agent which causes DNA or chromosome damage is termed:**

A mutagen.

❐❏ **The study of the relationship between factors that determine the frequency of distribution of disease within or between populations is termed:**

Epidemiology.

❐❏ **Twins who have all of their genes in common are termed:**

Monozygotic twins.

❐❏ **In order to perform paternity testing, samples are required from which family members?**

The child, the mother, and the putative father.

❐❏ **The most important genetic region in determining tissues for transplantation is that of the:**

Major histocompatibility complex genes (MHC).

❐❏ **Siblings usually have what percent of a chance of being HLA identical?**

25%.

❐❏ **If the carrier (heterozygote) frequency for a certain condition in the general population is 1/20 for a recessive condition, the incidence of the condition is:**

1/1600 - (1/20 x 1/20 x 1/4 = 1/1600).

❐❏ **In an extended pedigree with only male and female offspring of female parents being affected with a condition, with no offspring of males being affected, is most characteristic of:**

Mitochondrial inheritance.

❏❏ **In a family in which only males being affected through several generations which are apparently transmitted through females, the most likely mode of inheritance is:**

X-linked recessive inheritance.

❏❏ **If a man passes a genetic trait onto his son, what modes of inheritance can be ruled out?**

X-linked recessive and mitochondrial.

❏❏ **Why do abnormal recessive alleles not disappear from the population which undergoes random mating?**

Most alleles exist in the heterozygous form, carriers sometimes have or historically have had a selective advantage.

❏❏ **In parts of the world where malaria is endemic, the gene frequency for the Sickle Cell gene is higher than in areas where malaria has been eradicated. The reason for this is:**

Heterozygotes for Sickle Cell have increased fitness compared to non-carriers in areas where malaria is endemic.

❏❏ **The highest genetic risk for offspring of men over 45 years of age is that of:**

A new dominant mutation.

❏❏ **A healthy woman over the age of 35 has an increased risk for offspring, over that of the general population, for:**

Trisomies resulting from maternal nondisjunction.

❏❏ **Second degree relations (uncle, aunt, grandparent, nephew, niece, grandchild, half-sibling) have what proportion of genes in common with a proband?**

1/4.

❏❏ **Alternative forms of a given gene are termed:**

Alleles.

❏❏ **The term which describes different alleles at a specific locus is a:**

Heterozygote.

❏❏ **Monozygotic twins have what percentage of their genes in common?**

100%.

❐❑ **Family members who have 50% of their genes in common include:**

Parent, sibling, child, dizygotic twin - first degree relatives

❐❑ **This pedigree symbol indicates**

A consanguineous union

❐❑ **This pedigree symbol indicates**

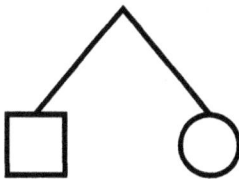

Dizygotic twins, male and female.

❐❑ **This pedigree symbol indicates**

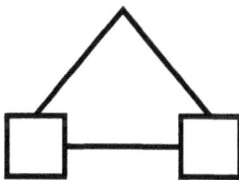

Monozygotic twins, males.

❐❑ **This pedigree symbol indicates**

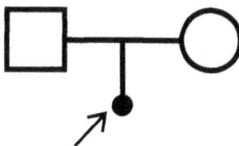

Miscarriage or abortion, sex not specified.

❐❑ **This pedigree symbol indicates**

Death of a female at 60 years of age.

❑❑ **This pedigree symbol indicates**

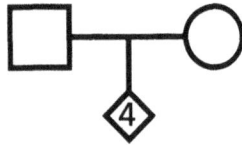

4 children, sex unspecified.

❑❑ **Genes which are closely located on the same chromosome are called:**

Linked.

❑❑ **Among the Mendelian patterns of inheritance, including autosomal dominant, autosomal recessive, X-linked recessive, and multifactorial, which is generally the most severe type of disorder?**

Autosomal recessive.

❑❑ **You have a couple in your practice who have had two children with achondroplasia, an autosomal dominant condition. Both parents are clinically normal. What is the most likely reason for this?**

Gonadal mosaicism.

❑❑ **You have a couple in your practice who are both deaf. Both have deaf siblings, but all of their parents have normal hearing. Your couple has 4 children who have normal hearing. This is explained by:**

Locus heterogeneity - each parent has a different recessive gene for deafness.

❑❑ **You have a 25-year-old male in your practice who has hemophilia B (X-linked recessive). You explain that his risks to offspring are:**

His sons will be unaffected, and his daughters will be carriers for the trait.

❑❑ **Two clinically normal parents have a child with cystic fibrosis. The probability that an unaffected child of this couple being a carrier is:**

2/3.

❑❑ **Two albino parents have a child with normal pigmentation. This is most likely explained by:**

Genetic heterogeneity (two distinct loci).

❑❑ **The practice of increasing the frequency of desirable genetic traits and decreasing the frequency of deleterious genetic traits is considered to be the phenomenon of:**

Eugenics.

❏❏ **The term used for conditions in which the recurrence risk for siblings with an affected child are lower than that for a Mendelian trait and when multiple genes are thought to be involved is that of:**

Multifactorial.

❏❏ **What type of genetic trait can be ruled out when a father transmits a specific disease to his son?**

Mitochondrial and X-linked.

❏❏ **When a person has an autosomal dominant gene which is not apparent, this is a sample of:**

Incomplete penetrance.

❏❏ **In a couple in which consanguinity is involved, the most likely mode of inheritance for a subsequent problem in offspring is that of:**

Autosomal recessive.

❏❏ **In autosomal dominant diseases when there is near 0 reproductive fitness, the most likely mechanism for the high frequency of these diseases is that of:**

New mutations.

❏❏ **If the frequency of an autosomal recessive disease is 1 in 6,400, what is the frequency of the carrier state in this population?**

1/40 (1/40 x 1/40 x 1/4 = 1/6400).

❏❏ **There are numerous lethal recessive genes which don't disappear from the population, including those which undergo random mating. Why is this?**

Most alleles exist in the heterozygous form and carriers my have (or had) a selective advantage.

❏❏ **You have a clinically normal patient in your office who has a sister with cystic fibrosis. What is his chance of being a carrier for this condition if he is clinically normal?**

2/3.

❏❏ **You have a patient referred to you because of an ultrasound evaluation which indicated a posterior encephalocele, polydactyly, large kidneys, and a probable cleft lip and palate. Chromosome studies revealed a 46,XY and she asks what her recurrence risk might be for a similar outcome in future pregnancies?**

Approximately 25%, since this is most likely an autosomal recessive condition which is characterized by Meckel syndrome.

❏❏ You have a couple in your office in which the husband has Waardenburg syndrome and his wife has Marfan syndrome (both are autosomal dominant disorders). What is a probability that their first child will be free of both of these conditions.

1/4.

❏❏ You have a couple in your practice in which the male has achondroplasia, an autosomal dominant disorder. His parents are unaffected but they ask about the risk for their offspring having the same condition as the father? You counsel them that the risk is approximately:

50%.

❏❏ A common limb deformation resulting from the lack of fetal movement, quite often caused by decreased amniotic fluid, is that of:

Club foot.

❏❏ Of all women referred for prenatal diagnosis, the most common category is that of:

Advanced maternal age.

❏❏ Your 30-year-old patient consults you regarding her concerns about her father having had hemophilia A (an X-linked condition) and is concerned about the risks for her newborn son having the same condition. You inform her that this risk is:

50%.

❏❏ A couple in your practice has had 3 unexplained early miscarriages. The incidence of a balanced chromosome rearrangement in either parent is approximately?

10%.

❏❏ Your patient consults you because she has a previous child with a meningomyelocele and she has three other normal children. She asks about the risks for the next child having a neural tube defect. You respond that her risk is approximately:

3-5%.

❏❏ Your 30-year-old patient consults you because she has two sons with Duchenne muscular dystrophy (X-linked recessive). There is no other family history and she asks what the likelihood is that she is a carrier for this gene. Your response is:

Nearly 100%.

❐❐ If approximately 1 in 20 individuals in a population carry the gene for cystic fibrosis, what is the chance that two unrelated individuals will have a child with this disease?

1/600 (1/20 x 1/20 x 1/4).

❐❐ You have a couple in your practice in which both have Marfan syndrome, an autosomal dominant disease, and they ask what the chance for their offspring being affected with the same condition. You inform them that the risk is approximately:

75%.

❐❐ The recurrence risk for Down syndrome in offspring of a parent who has a t(21;21) balanced Robertsonian translocation is:

100%.

❐❐ The recurrence risk for Down syndrome for a parent with a 14;21 translocation is, for the father 10%, for the mother 3%. The recurrence risk for Down syndrome when both parents have a normal karyotype is approximately:

1%.

❐❐ When there is male to male transmission of a trait, this is most likely a mode of inheritance considered to be:

Autosomal dominant, also more rarely Y-linked inheritance.

❐❐ Your 20-year-old patient reports that her father has hemophilia B, an X-linked disorder. What is her chance of being a carrier?

100%.

❐❐ Your female patient reports that she has cystic fibrosis and she also has recently married a man who is a carrier for the same condition. What is their risk for having an affected child?

50%.

❐❐ A condition which appears in multiple generations of a family which has also been apparent in both sexes most likely follows what inheritance pattern?

Autosomal dominant.

❐❐ The approximate percentage of children born with a serious birth defect is approximately:

3%.

❏❏ **The most prevalent autosomal recessive gene in the eastern European population is:**

Tay Sachs (1/20).

❏❏ **The most prevalent autosomal recessive gene in the general Caucasian population is:**

Cystic fibrosis (1/25).

❏❏ **The most prevalent autosomal recessive gene in the general Mediterranean population is:**

Thalassemia (1/20).

❏❏ **The most prevalent autosomal recessive gene in the African American population is:**

Sickle Cell (1/10).

❏❏ **A newly married young couple consults you about whether any testing should be conducted before they begin a family. There is no reported family history of mental retardation or other concerns but both are of eastern European ancestry. A screening consideration for this couple would be for that of:**

Tay Sachs / Gaucher disease carrier state.

❏❏ **A couple consults you in which the wife has Waardenburg syndrome (autosomal dominant) and the husband has achondroplasia (autosomal dominant). What is the chance that their first child will have neither condition?**

1/4.

❏❏ **You have a couple in your practice who both have an AB blood type. What blood type can any offspring not have?**

O - all offspring will have AA, AB or BB.

❏❏ **You consult with a couple in which the wife has an AB blood type and the husband has an O blood type. What blood type can any offspring not have?**

AB or O - offspring will have A or B.

❏❏ **An X-linked recessive trait has a frequency of 1 in 500 males being affected. What is the carrier frequency among females?**

1 in 250.

❏❏ **Approximately how many single gene diseases (inherited in an autosomal dominant, autosomal recessive, or X-linked fashion) have been described?**

Approximately 4,000.

❏❏ **Sickle Cell anemia affects approximately how many Blacks in the United States?**

1/400 with approximately 1/10 being heterozygous or having the Sickle Cell trait.

❏❏ **What is the frequency of cystic fibrosis (in the homozygous state) among Caucasians in the United States?**

Approximately 1/2,000 with approximately 1/25 being heterozygotes for a CF gene.

❏❏ **Those traits which result from the interplay of multiple environmental factors with multiple genes is termed:**

Multifactorial.

❏❏ **Approximately how many functional genes have been thus far mapped to the X chromosome?**

Approximately 300.

❏❏ **It is estimated that there are approximately how many human genes per cell which will soon be placed in the genetic map upon completion of the Human Genome Project?**

Approximately 100,000.

❏❏ **The program which represents the largest research investment in ethics as it relates to the Human Genome Project is referred to as:**

ELSI (ethical, legal and social implications) program.

❏❏ **The frequency with which a test yields a positive result when the disease is present is termed:**

Sensitivity.

❏❏ **The frequency with which a specific test is negative when the disease is absent or the proportion of unaffected individuals who have a negative test is termed:**

Specificity.

❏❏ **The proportion of all false-positive test results occurring in individuals who are not affected is termed to be the:**

False-positive rate.

❏❏ **The proportion of all negative test results in individuals affected with a particular disease or trait is considered to be the:**

False-negative rate.

❏❏ **The occurrence of a genetic disease at an earlier age of onset or with increasing severity in successive generations is termed:**

Anticipation.

❏❏ **Random fluctuations in gene frequencies which are most evident in small populations is considered to be:**

Genetic drift.

❏❏ **It has been estimated that most humans carry approximately how many abnormal recessive genes in the heterozygotic state:**

4-8.

❏❏ **Duchenne Muscular Dystrophy has a much higher mutation rate than does Huntington's disease. Why is that?**

The gene for Duchenne Muscular Dystrophy is much larger than is the gene for Huntington's disease, thereby giving a greater chance of mutation.

❏❏ **What criteria are utilized to determine the efficacy of newborn screening?**

The test must be accurate and treatment must be available.

❏❏ **The study of the genetic basis for differences in response to drugs is termed:**

Pharmacogenetics.

❏❏ **The complex of human leukocyte antigen (HLA) genes is called:**

Major histocompatibility complex (MHC).

❏❏ **The A, B, and O genes are alleles mapped to chromosome 9q34. Of the A, B or O genes, which is recessive:**

O.

❏❏ **A dramatic adverse response to anesthetic drugs particularly halothane and muscle relaxants and which is characterized by high fever, muscle rigidity and hypercatabolism caused by an anesthetic disturbance to calcium metabolism is termed:**

Malignant hyperthermia.

❏❏ **The mode of inheritance in which the more severe the malformation the higher the recurrence risk is that of:**

Multifactorial inheritance.

❏❏ **The conditions of cleft lip / palate, pyloric stenosis and neural tube defects, are considered have what type of etiology?**

Multifactorial.

❏❏ **The term which describes the wide spectrum of phenotypic changes resulting from a single genetic alteration, where many systems are affected due to a mutation in one critical protein is that of:**

Pleiotropy.

❏❏ **In genetic testing, the term used to describe the proportion of all normal individuals who have normal test results is termed:**

Specificity.

❏❏ **In genetic testing, the proportion of all affected individuals who have abnormal test results is termed:**

Sensitivity.

❏❏ **In performing a Chi square analysis, what level of probability is considered to be significant?**

Less than 0.05.

❏❏ **The type of genetic analysis which involves analyzing the relative probability of two alternatives is that of:**

Bayesian analysis.

❏❏ **Most inborn errors of metabolism are characterized by having what Mendelian inheritance pattern?**

Autosomal recessive.

❏❏ **The heterozygotic state of cystic fibrosis is thought to provide some immunity to what disease?**

Tuberculosis.

❏❏ **The heterozygotic state of Sickle Cell is thought to provide protection against what disease?**

Malaria.

BIBLIOGRAPHY

BOOKS / ARTICLES

Gardner, R.J.M. and Sutherland, G.R. Chromosome Abnormalities and Genetic Counseling, *Oxford Monographs on Medical Genetics No.29* (2nd edition), Oxford University Press, New York, NY, 1996.

Heim, S. and Mitelman, F. Chromosomal and Molecular Genetic Aberrations of Tumor Cells. *Cancer Cytogenetics* (2nd edition) Wiley Publication, New York, NY, 1995.

Jones, K.L. *Smith's Recognizable Patterns of Hman Malformation*, 5th edition, W. B. Saunders Company, 1997.

Jorde, L.B. et al., *Medical Genetics*, 2nd edition, Mosby, St. Louis, MO, 2000.

Rimoin, D.L. et al. (Ed.) *Principles and Practices of Medical Genetics*, 3rd edition, Churchill Livingstone, 1997.

Rooney, D.E. and Czepulkowski, B.H. (Ed.) A Practical Approach: Volume I Constitutional Analysis, *Human Cytogenetics* (2nd edition), Oxford University Press, New York, NY, 1992.

Rooney, D.E. and Czepulkowski, B.H. (Ed.) A Practical Approach: Volume II Malignancy and Acquired Abnormalities, *Human Cytogenetics* (2nd edition), Oxford University Press, New York, NY, 1992.

Sack, Jr., G.H. *Medical Genetics*, McGraw-Hill Health Professions Division, 1999.

Sandberg, A.A. *The Chromosomes in Human Cancer and Leukemia* (2nd edition), Elsevier Science Publishing, New York, NY, 1990.

Verma, R.S. and Babu, A. *Human Chromosomes: Principles and Techniques*, (2nd edition), McGraw-Hill, 1994.